Dynamics of Innovation

Dynamics of Innovation

The Expansion of Technology in Modern Times

François Caron

Translated and Abridged by

Allan Mitchell

Published in 2013 by
Berghahn Books
www.berghahnbooks.com

English-language edition
© 2013, 2016 Berghahn Books
First paperback edition published in 2016

French-language edition
© 2010 Editions Gallimard, Paris
La dynamique de l'innovation
By François Caron

All rights reserved.
Except for the quotation of short passages
for the purposes of criticism and review, no part of this book
may be reproduced in any form or by any means, electronic or
mechanical, including photocopying, recording, or any information
storage and retrieval system now known or to be invented,
without written permission of the publisher.

Library of Congress Cataloging-in-Publication Data

Caron, François, 1931–
  [Dynamique de l'innovation. English]
  Dynamics of innovation : the expansion of technology in modern times /
    François Caron ; translated and abridged by Allan Mitchell.
    p. cm.
  Includes bibliographical references and index.
  ISBN 978-0-85745-723-3 (hardback) ISBN 978-1-78533-036-0 (paperback)
    1. Technological innovations—Europe—History. 2. Technological
  innovations—Economic aspects—History. 3. Technological innovations—Social
  aspects—History. 4. Technological innovations—History. 5. Knowledge, Theory
  of—History. I. Mitchell, Allan, 1933– II. Title.

HC79.T4C39613 2012
338'.064—dc23

2012001643

British Library Cataloguing in Publication Data
A catalogue record for this book is available from
the British Library.

ISBN 978-0-85745-723-3 (hardback)
ISBN 978-1-78533-036-0 (paperback)

Contents

List of Abbreviations	viii
Translator's Preface	xii
Introduction	xv

Part One   1

Chapter 1: The Artisanal Mode of Knowledge   5
    Industrial Framework: The World of Trades, 6
    Artisanal Knowledge and the Arc of Experience, 9
    The Interaction of Trades, 16
    The Circulation of Knowledge, 21

Chapter 2: From Artisan to Expert   23
    The Appropriation of Artisanal Knowledge, 24
    The Role of Writing, 25
    Knowledge of Experts, 27

Chapter 3: Formalized Knowledge   31
    The Professional Engineer, 31
    The Industrial Entrepreneur, 34
    Science and Utility: The Other Revolution, 35
    Practice and Theory, 39

Chapter 4: Technological Adventures   41
    The Production of Energy, 41
    The Mechanization of Industry, 43
    The Birth of the Mechanical Industry, 46
    The Chemical Industry, 47

Part Two   49

Chapter 5: Institutional Logic and the Dynamics of Knowledge   51
Entrepreneurs and Enterprises, 52
Artisanal Trades and the Formalization of Knowledge, 62
Engineers and Engineering, 64
A Global Model of Constructing Engineering Knowledge, 70
The Development of Engineering Science: Three Examples, 72
Science, the Universities, and the State, 79

Chapter 6: Steam Engines   84
Domination of Empirical Knowledge before 1850, 85
The Birth of Thermodynamics, 90
Experimental Thermodynamics, 96
The Conquest of Great Efficiency, 98
The End of an Era, 103

Chapter 7: The Chemical Industry   105
Organic Chemistry and the Dye Industry before 1900, 106
Physical Chemistry in the Second Industrial Revolution, 118
Macromolecular Chemistry and Vertical Integration, 130
Summing Up Chemistry, 140

Part Three   145

Chapter 8: Technological Interdependence and Consumer Needs   147
Iron Metallurgy in France in the Nineteenth Century, 148
Generalizing the Model, 155

Chapter 9: Strategies and Social Networks   160
Global Communities: Gas, Electricity, Automobiles, 160
Social Groups, 162
Enterprises and Networks, 166
The Pillars of Innovation, 172
Local Productive Systems, 177

Part Four   179

Chapter 10: From Early Modern Times to the 1880s   181
The English Model, 181
The Rise of Mass Civilization in Paris, 1830–1880, 183

Chapter 11: Technological Networks and Communications　　　193
French Railways: Rationalization and Cybernetics, 193
Interconnections: Networks of Electricity, 195
Mass Consumption, 198
Mass Production, 199
The Rise of Communications, 200
Telecommunications, 201
The Birth of a Communications Society, 203

Chapter 12: From Microprocessors to the Internet　　　205
Flexible Production, 206
Telecommunications, 207
Information and Audiovisual Technologies, 208
Computer Networks and the Birth of the Internet, 209
Communications of a Large Network: The SNCF, 211
The Second Life of Networks, 1995–2008, 211

Chapter 13: Information Technologies and Society　　　213
Enterprises, 213
Objects of Daily Life, 216
The Era of the Internet and Cellular Phones, 219
The Nature of Messages, 220
Social Connections, 220
The Unforeseen Outcome, 222

Conclusion　　　223
Notes　　　231
Selected Bibliography　　　243
Name Index　　　249

Abbreviations

ABB	Asea Brown Boveri
ADR	Association pour le Développement de la Recherche
ADSL	Asymmetric Digital Subscriber Line
AEG	Allgemeine Elektrizitäts-Gesellschaft
Anvar	Agence nationale de valorisation de la recherche
APT	Automatically Programmed Tool
ARPA	Advanced Research Project Agency
Artemis	Atelier de recherches sur les techniques mathématiques et informatiques des systèmes
ASCII	American Standard Code Information Interchange
ASME	American Society of Mechanical Engineers
ASTM	American Society for Testing Materials
AT&T	American Telephone & Telegraph
BASF	Badische Anilin- und Soda-Fabrik
CABT	Courant Alternatif Basse Tension
CAHT	Courant Alternatif Haute Tension
CAO	Conception Assistée par Ordinateur
CB	Citizen Band
CCBT	Courant Continu Basse Tension
CCHT	Courant Continu Haute Tension
CCIF	Comité Consultatif International de la Téléphonie
CCIT	Comité Consultatif International de la Télégraphie
CCITT	Comité Consultatif International Télégraphique et Téléphonique
CD	Compact Disc
CEA	Commissariat à l'Énergie Atomique
CENG	Centre d'Études Nucléaires de Grenoble
CEPT	Conférence Européenne des Postes et Télécommunications
CERN	Conseil Européen pour la Recherche Nucléaire
CET	Common Efficiency Team
CFAD	Compagnie Française de l'Acétylène Dissous

CGE	Compagnie Générale d'Électricité
CGS	Centimètre, Gramme, Seconde
CII	Compagnie Internationale pour l'Informatique
CII-HB	CII Honeywell-Bull
CNET	Centre National d'Études des Télécommunications
CNAM	Conservatoire National des Arts et Métiers
CNRS	Centre National de la Recherche Scientifique
CPDE	Compagnie Parisienne de Distribution de l'Électricité
CP/M	Control Program for Microcomputers
CSF	Compagnie Générale de la Télégraphie Sans Fil
Datar	Délégation à l'aménagement du territoire et à l'action régionale
DETE	Direction des Études de Traction Électrique
DGA	Division Générale de l'Armement
DGRST	Direction Générale de la Recherche Scientifique et Technique
DOS	Disk Operating System
DVA	Divinyl Acétylène
DVD	Digital Video Disc
EDF	Électricité de France
EDVAC	Electronic Discrete Variable Automatic Computer
ENIAC	Electronic Numerical Integrator and Computer
ENS	École Normale Supérieure
ESSA	École Supérieure du Soudage et de ses Applications
FAO	Fabrication Assistée par Ordinateur
FCC	Federal Communications Commission
FM	Frequency Modulation
FMS	Flexible Manufacturing System
GCTM	Gestion Centralisée du Trafic Marchandises
GE	General Electric
GEM	General Electric Metallized Lamp
GPS	Global Positioning System
GSM	Global System for Mobile Communications
HP	Hewlett-Packard
HT	High Tension
IAS	Intelligent Autonomous System
IATM	International Association for Testing Material
IBM	International Business Machines
ICI	Imperial Chemical Industry
IEC	International Electrotechnical Commission
IETF	Internet Engineering Task Force
IGEC	International General Electric Company
IMAG	Institut d'Informatique et Matématiques Appliquées de Grenoble
INPG	Institut National Polytechnique de Grenoble

IPG	Institut Polytechnique de Grenoble
IRSID	Institut de Recherche de la Sidérurgie
ISO	International Standardization Organization
JVC	Japan Victor Company
LAG	Laboratoire d'Automatique de Grenoble
LEM	Laboratoire d'Essais Mécaniques des Métaux, Chaux et Ciments
LEPM	Laboratoire d'Électrostatique et de Physique du Métal
LETI	Laboratoire d'Électronique et de Technologie de l'Information
LHD	Laboratoire Dauphinois d'Hydraulique
LSTS	Large-Scale Technological System
MIC	Modulation par Impulsion et Codage
MIT	Massachusetts Institute of Technology
MITI	Ministry of International Trade and Industry
MS-DOS	Microsoft Disk Operating System
NASA	National Aeronautics and Space Administration
NAW	Nouvel Acheminement des Wagons
NBPP	Neyret-Beylier et Piccard-Pictet
OCA	Office Central de l'Acétylène
Onera	Office national d'études et de recherches aérospatiales
OS	Operating System
PC	Personal Computer
PIB	Produit Intérieur Brut
PME	Petites et Moyennes Entreprises
POS	Point-of-Sale
PVC	Polyvinyl Chloride
RAM	Random-Access Memory
RESA	Réservations et Suppléments Associés
RETIF	Réseau de Téléinformatique Ferroviaire
Retipac	Réseau de téléinformatique à commutation de paquets
RNIS	Réseau Numérique à Intégration de Services
ROM	Read-Only Memory
RTE	Réseau Techno-Économique
SACM	Société Alsacienne de Constructions Mécaniques
SADSE	Société d'Acétylène Dissous du Sud-Est
SAF	Soudure Autogène Française
SAGE	Semi-Automatic Ground Environment
SAPPHO	Scientific Activity Predictor from Patterns with Heuristic Origins
SBU	Strategic Business Unit
SDET	Société de Développement de l'Enseignement Technique
SEA	Société d'Électronique et d'Automatique
SECEM	Société d'Électro-Chimie et d'Électro-Métallurgie

SEMF	Société Électro-Métallurgique Française
SEMS	Société Européenne de Mini-informatique et Systèmes
SESSIA	Société d'Études et de Construction de Souffleries, Simulateurs et Instrumentations Aérodynamiques
SFE	Société des Chemins de Fer de l'Est
SFR	Société Française de Radioélectricité
SIE	Société Internationale des Électriciens
SNCF	Société Nationale des Chemins de Fer Français
Sogreah	Société grenobloise d'études et d'applications hydrauliques
SPL	Système Productif Local
TBU	Technical Business Unit
TCP/IP	Transfer Control Protocol/Internet Protocol
TGV	Train à Grande Vitesse
UIT	Union Internationale des Télécommunications
UIT-T	Union Internationale des Télécommunications, secteur de la normalisation des Télécommunications
Upepo	Union des producteurs d'électricité des Pyrénées-Occidentales
USE	Union des Syndicats d'Électricité
UTI	Union Télégraphique Internationale
UUCP	Unix-to-Unix Copy Protocol
VHS	Video Home System
ZIRST	Zone pour l'Innovation et les Réalisations Scientifiques et Techniques

Translator's Preface

François Caron is France's leading historian of technology. As a scholar, he is undoubtedly best known for his remarkable studies of the French railway system. But he has also done significant work on the history of electrical, water, and steam power, the theory of innovation, the structure of enterprise, and various other aspects of economic development in the nineteenth and twentieth centuries. In this volume, he combines these different facets of his research in order to present a broad panorama of modern technology.

The book opens with a complex theoretical analysis of different types of knowledge and the means of their transmission. The first hero of this account is the artisan who, having acquired his skills through apprenticeship under a master, intuitively performs a trade by repeating identical "gestures" that may be adapted to particular tasks. This is what Caron calls "artisanal know-how." With time, these procedures are observed by experts, engineers, intellectuals, and academics who begin to generalize about them. There is thus a natural evolution toward the theorization and codification of the artisan's experience—in a word, science. Know-how thereby becomes formalized knowledge.

To illustrate this abstract scenario, Professor Caron rolls out an ambitious agenda covering several centuries and four major national contexts: Great Britain, France, Germany, and the United States. After an introductory sketch of medieval and early modern developments, he presents a story of the three industrial revolutions of the modern era. First, Britain erupted in the early nineteenth century with the new textile industry, the use of steam power, and the launching of railroads. Then, in the late nineteenth century, Germany took the lead in crucial industries, such as dyes, electronics, and automobiles. Finally, it was the turn of the United States to advance by developing computers and the Internet in the latter half of the twentieth century.

This sprawling and yet quite orderly narrative displays three distinguishing characteristics. First of all, Caron's extraordinarily versatile expertise enables him to view a host of technological phenomena in a broad perspective. He has left hardly a stone unturned: tools and trades, pottery and clocks, watermills and windmills, printing and astronomy, elegant furniture and coal mining, mechanics and metallurgy, steam engines and textiles, highways and railways, organic chemistry and electricity, turbines and thermodynamics, polymers and plastics, autos and airplanes, computers and copying machines, cell phones and the Internet. Specialists of certain areas, periods, or disciplines may find his treatment of their domain familiar, but very few readers will possess a grasp of so many aspects of technological growth. Whatever our specialty or the general state of our understanding, we all have much to learn and to discover in this book.

A second conspicuous feature is the attention to chronology. Caron takes pains to be precise about when specific inventions or innovative procedures first appeared. The progress of technological patterns was manifestly complex, interconnected, and often simultaneous. To establish the where, what, and when of innovation is accordingly basic. Caron does not ignore nuances or skimp on detail while explaining how dynamic modern industries and international networks were created.

Third, whereas Professor Caron certainly gives Britain, Germany, and America their due, he also makes abundant references to France, and he frequently provides French examples to make his point. It is fair to say that he writes with a French accent. In doing so, he succeeds in showing how French artisans, engineers, entrepreneurs, and scientists made constant and important contributions to technological change. Therewith he offers a valuable addition to the scholarly literature that sometimes tends to ignore or to underestimate France's role in the industrialization of Europe since the days of Napoleon Bonaparte.

Translating this text into simple and comprehensible English presents several problems. Caron is the heir of a rich French rhetorical tradition stretching back at least to Michelet and Tocqueville. Hence, he has a tendency to pack as much as possible into a single sentence, which in English must on occasion be broken down into two or three. Sentences that flow in French may sound quaint or stilted in another tongue, a challenge well known to translators of Balzac, Flaubert, or Proust. When articles are plural in French, they are better left singular in English, or vice versa. Moreover, some words cannot be exactly translated: *savoir*, for instance, has been rendered either as 'knowledge', 'learning', or 'understanding', depending on the context. It is noteworthy that the initially prominent distinction between the *savoir-faire* of the artisan and the *savoir* of academics and intellectuals begins to fade as technology and science gain an ever more intimate proximity. Despite these nagging difficulties, every effort has been made to retain the tone, the rhythm, and above all the sense of Professor Caron's prose.

It is well to add a final word about abridgments. They have generally been of two sorts. One occurs in an attempt to simplify a sentence in order to draw out its essential meaning, an unavoidable result of the difference in linguistic styles. The second involves the omission of some passages or paragraphs that seem exceedingly detailed or technical and that therefore contain material of only marginal interest for most anglophone readers. Specialists in this or that topic under discussion will of course want to consult Professor Caron's original French text. It is to be hoped that nothing fundamental for the general reader has been excised. In short, the translator's purpose must be, and has been, to enable others to enjoy and appreciate this learned and splendid work of history.

— *Allan Mitchell*

Introduction

Technology has invaded our daily existence as well as that of the state and of society. This ubiquity prompts us to inquire about the reasons for the transformation of technological knowledge. Numerous theories have been proposed to account for the development and the complexity of the interactions that occur in this perpetually changing aspect of civilization. In order to illuminate them, this volume offers a discussion centered on a classical typology of knowledge and on a chronology of its evolution in Europe from the waning of the Middle Ages to the beginning of the twenty-first century.

We can distinguish three types of technological learning:

- *Tacit knowledge* may be manual, olfactory, auditory, visual, or even graphic and seemingly belongs to "know-how." Its transmission is realized through gesture, speech, or graphic design.
- *Formalized knowledge* is descriptive and discursive. It is constructed from tacit knowledge combined with actual experience in an effort to formulate its theoretical contents. It can also be the result of a positive effect of scientific learning, thereby explaining know-how with reference to "know-why."
- *Codified knowledge* is the result of the natural evolution of formalized knowledge through oral transmission in the form of conferences or through the publication of treatises, articles, and books.

These types of knowledge may assume a purely empirical character as the fruit of accumulated experience, or they may be the result of an informed analysis of practical gestures, or they may emerge from the transformation of technology by a scientific knowledge oriented from the beginning toward a search for truth rather than utility.

These three modes of intellectual construction naturally cohabit today at the core of technological knowledge. There is a long-standing tendency to diminish the pre-eminence of tacit knowledge in favor of formalized knowledge, and then of formalized knowledge in favor of codified knowledge, albeit without causing the disappearance of the first two.

A historical approach seems to be best suited to seize the original character of each of these three evolving types of knowledge, which must be embedded in historical reality in order to take shape. The four parts of this work illustrate the different epochs of this transformation in Europe between the twelfth and the twentieth centuries. Two long phases are clearly apparent: the first extended from the twelfth century to the opening of the nineteenth; the second ran from the 1830s to the end of the twentieth century. The former ended with the emergence between 1760 and 1830 of the first industrial revolution, which opened the way for the expansion of another model of acquired knowledge between 1830 and 1960. This latter model has in turn undergone a radical transformation between 1970 and 2000, which appears to mark the birth of another long epoch in the history of technology.

During the course of the first two periods identified above, the structures of knowledge and technological change have become products of a complex interaction among four elements in a global system: (1) the formation of a corps of technicians and scientists organized into specific groups; (2) the development of powerful institutions devoted uniquely to scientific and technological research; (3) the creation of a network of social relations among the many actors in structuring knowledge; and (4) the establishment of relations between the consumption of knowledge and the various modes of production.

Part One of this work is devoted to a comprehensive analysis of the evolution of these key components from the twelfth century to the beginning of the nineteenth century. Reviewed within this framework are the transformations that caused Europe to pass from a technological system characterized by the slow evolution of practical techniques on the basis of tacit knowledge to a system founded on formalized and codified knowledge. The principal actors in this evolution were artisans, many of whom were grouped into guilds. A transformation occurred in the interior of these institutions, allowing the spread of artisanal skills into the realm of experts and engineers. At the same time, scientific knowledge penetrated several disciplines, and a voluntary orientation toward technological education was put into place by the modern state. These developments resulted between 1760 and 1830 in the emergence and dissemination of the first industrial revolution.

Part Two concentrates on the impressive groups of technicians and scientists constituted from the debut of the nineteenth century to the Second World War. Some of these groups, such as industrial mechanics and hydrology, are examined, as well as two long-term trajectories: the fixed steam engine during the nineteenth century and organic chemistry between 1850 and 1930. This section underscores the spectacular rise of research institutions within enterprises and universities. These became sites of confrontation of both complementary and competing visions of intellectual progress and thus were the most conspicuous forms of the institutionalization of research.

Part Three emphasizes the social relations of work and their impact on the construction of knowledge. After a brief excursion into the history of medieval and modern metallurgy, the period between 1830 and 2000 is highlighted in order to identify the social roots of innovation in a technological society ceaselessly renewed by invention. This analysis distinguishes various concepts of interdependence among groups dedicated to the constant adaptation of technology to needs and to social construction, thereby exploring the vast terrain of relationships between customer and producer, between client and purveyor, between consumer and entrepreneur, between marketing and research.

Part Four, with its point of departure in Great Britain in the eighteenth century, analyzes the relations established between consumers and modes of production, which were all part of technological networks on a gigantic scale that enjoyed a remarkable takeoff after the 1830s and even more after 1880. These technological networks had a binary character related to particular problems raised by their control of large territories and by their domination over all forms of social activity and collective endeavor.

These various analyses should permit a response to many of the questions posed by the development of a technological society, including the following:

- What are the determinants of technological choice adopted by engineers, researchers, and managers to orient globally technological change, such as electric lighting and, in particular, the incandescent lamp?
- Given the enterprise as a privileged means of convergence and construction of knowledge, what has been the evolution of its organizational model since the eighteenth century in relation to available technology, and what have been the consequences?
- What is the role of relations established by enterprises with other actors in technological change? In other words, what is the role of social networks in the development of enterprises and in technological change?
- How are different forms of knowledge differentiated and constituted to create a dynamic of growth and innovation? This question comprises three: How was the transition achieved from artisanal knowledge to an engineering knowledge and from there to a knowledge strongly impregnated with scientific learning? What was the role of institutions of teaching and research in this process? And what was the role of the internal dynamics of transformation in scientific knowledge and technological practice?
- What remains of the concept of an industrial revolution? Has it lost its relevance when the history of technology suggests a process of continuity rather than a rupture and when it also allows a valid distinction among three separate industrial revolutions: 1760–1830, 1880–1914, and 1970–2000?

The conclusion should bring together the answers to questions such as these.

Part One

Technological knowledge may be tacit or formalized. In the latter instance, it can also be codified. Tacit knowledge may be defined as a learning that is not available to those who possess it in a discursive or theoretical form but which appears instead as "know-how." It allows the accomplishment of tasks, such as cycling or controlling the fusion of metals, without acquiring the explicit knowledge of the facts and causal relationships that make them possible and effective. Tacit knowledge is enveloped in the culture and experience of the actor. It can exploit the actor's glance, touch, or sense of smell and requires a perfect mastery of the gestures necessary for the completion of an action, but it is transmitted only within informal exchanges by means of imitation or eventually of demonstrations and graphic illustrations.

Formalized knowledge is descriptive and discursive. It is communicated by words, written works, and graphics. It may be explicit, and it tends to theory and science. It may become the object of codification and emerge through the definition of rules, which are no more than simple recipes or explicit hypotheses. An engineering science is thus created, transmitted by oral or written means, and elaborated by expert analyses circulated in countless tracts and textbooks made available to all.

Formalized and codified knowledge may have an empirical character, since the concept of empirical knowledge is different from that of tacit knowledge, although these two notions are often falsely confused. Empirical knowledge is the result of new technological constructions formed in the unique professional experience of apprenticeship, which may assume an intergenerational character. The empirical concept does not describe the state of knowledge but rather its manner of creation. It constitutes an arc of experience of which economists have stressed the positive effects on the growth of productivity. Hence, empirical knowledge may have a discursive or formalized character that makes possible its oral, written, or graphic transmission.

Starting with a given state of learning, technological change is the product of practical innovations, creating new knowledge that grafts itself onto existing knowledge or substitutes for it. This substitution triggers a process of reciprocal interaction at the core of the system, which provokes a recombination of practice and a reorganization of knowledge. In 1967 Fernand Braudel wrote: "Everything has always been technological, violent action but also the patient and monotonous efforts of men modeling a stone or a piece of wood to make tools or weapons. Is that not a sweeping activity, essentially conservative, slow to evolve, which science (its belated superstructure) gradually acquires, if it is acquired at all?"[1] Despite this praise of gradualness, it is necessary—in order to explain the emergence of huge concentrated zones of technological method in eleventh-century Europe, fifteenth-century Italy, seventeenth-century Holland, or eighteenth-century England—to invoke an external factor, namely, science. Yet even if this factor played an important role in each of those "industrial revolutions," it cannot explain everything, since the slowness of technological evolution did not limit its capacity to effect radical change. As John Harris has written, "the gradual development of a process may have revolutionary effects in the sense of a radical transformation of technology and the economy."[2]

This gradual dynamic of know-how has been the dominant trait of the history of technology from the eleventh to the nineteenth century. In this space of time, innovations more often resulted from a process of collective apprenticeship than from ingenious inventors. Their role, however, became increasingly frequent after the thirteenth century. Thus, in 1474 the Venetian Senate justified publication of a text that established the legal basis of patents for invention: "Eminent persons are among us who are capable of inventing ingenious procedures and, every day, of augmenting the grandeur and power of our city. If we adopt measures to protect these procedures against imitation, which deprives the inventor of honor, we would inspire new talents that will be of the greatest utility to our community."[3] This text shows that invention was perceived as the fruit of clearly identified individual capacity and not as a collective activity. Taking the contrary view, liberal authors such as Michel Chevalier maintained in the nineteenth century that the inventor merely added a final touch to the totality of research in progress.

These different concepts seem capable of illuminating the process of constructing technological knowledge in Europe since the twelfth century by distinguishing three modes of knowledge that could be successfully unified in a global system. The first dominated until the beginning of the nineteenth century. It was marked by the development of artisanal knowledge and the progressive transformation of much of it into formalized learning, thereby constituting a coherent corpus of learning on which the technologies of the industrial revolutions could be elaborated. The second was influenced by the development of engineering sciences, which emerged progressively from the Middle Ages to the nineteenth century. The third took flight after the seventeenth century and triumphed at the end of the nineteenth century, resting as it did on a permanent dialogue between technological practice and scientific knowledge.

The confrontation among the possessors of these three types of knowledge—the empirical knowledge of artisans, the formalized knowledge of engineers, and the scientific knowledge of intellectuals—bore increasingly plentiful fruit in the domain of learning and invention since the eighteenth century, which became intensified and generalized in the course of the nineteenth century.

Chapter 1

The Artisanal Mode of Knowledge

The initial episode in the process of the construction of knowledge was its application, not only in the artisan's workshop or domestic employment, but also in large work projects or in manufacturing during the early modern age. In such a system, both tacit and empirical knowledge were incorporated in the gestures, the glances, the touch, or the sense of smell of operators, artisans, and workers. It was first of all a manual skill—a kind of "knowledge of the hand." Its verbal, written, or graphic expression (by means of drawings or models) conformed to rules that produced not causal analyses or mathematical laws but simple recipes. This knowledge was transmitted in a hereditary fashion by apprenticeship or the circulation of workers from one workshop or construction site to another. It was not immutable but rather in constant evolution. In time, this knowledge became increasingly complex and sophisticated while still remaining to a great extent tacit. It was also an empirical knowledge that could be a source of innovations and consequently a creator of fresh knowledge.

At the end of the eleventh century, the production of manufactured objects was assured above all within a seigniorial framework by a largely servile labor force. It was founded on knowledge and know-how that had been inherited principally from the Roman Empire, either directly or through resurgence during the Carolingian era, although they had also been subjected to "barbarian" or Oriental influences. This sum of knowledge had suffered important losses in numerous realms. In the building trades, for instance, the quantity of tools deployed was considerably reduced since the Roman era, and the beautiful edifices of cut stone had disappeared. Nonetheless, in other crafts, such as gold-plating or armaments, refined know-how had been protected and preserved, at least in some centers of excellence.

Thanks to rural and urban artisans, between the eleventh and thirteenth centuries a corpus of coherent and complex knowledge was composed in all sectors of activity by individuals working independently, often with the aide of valets and apprentices. After the twelfth century, another organization of work appeared that was founded on the integration of artisans and domestic workers into enterprises controlled by merchant-manufacturers who retailed finished products in local or distant markets. The place of work might be rural, urban, or a combination of both. This model realized a considerable expansion before contracting during the second third of the nineteenth century, yet without completely disappearing. Within the framework of this commercial capitalism, certain activities, such as preparing the fabrication of cloth, might employ several dozens of workers. In England and on the Continent, many establishments directed by industrialists existed in sectors like glass and paper or in the production of non-ferrous materials. They have been considered by certain economic historians as true industrial enterprises. In reality, however, these proto-industries were only "rare birds," remarks François Crouzet.[1] Likewise, the state manufacturers created in France during the seventeenth century were rarely large, concentrated enterprises, and one cannot yet speak of factories in their regard, even if the goal was to improve the execution of labor. For the most part, they used, within a limited space, the technology developed by artisans or domestic workers. The best organized of these "factories" prepared the transition to a progressive adaptation of a rational and continuous productive process. But the dynamism of the medieval and early modern economy rested mainly on artisanal work, even when it was sometimes executed in large enterprises.

Industrial Framework: The World of Trades

It was within this general framework that a fund of knowledge was constituted after the twelfth century that splintered into more and more specialized trades with increasingly limited tasks. These trades became organized into variously named communities that were designated as guilds after the thirteenth century. They were subjected to regulations that defined their spheres of operation, the characteristics of their products, the nature of authorized procedures, and the punishment for infractions. Urban artisans were subjected to much stricter rules than those of rural artisans, who were sometimes exempted altogether. The express purpose of the rules was to guarantee the quality of production. They might stipulate, for example, a ban on undertaking activities other than those defining a trade, a limitation on the number of apprentices that the master had a right to engage, or a prohibition of competition among masters. Such provisions resulted in monopolies, the exclusivity of trades, the hereditary transmission of a master's status, and the retention of secretive methods of production. In early modern times, the power of guilds was maintained, but in a country like Great Britain the influence of guilds declined steeply in the eighteenth century. Guild regulations in many cases hindered the creation of fresh knowledge by slowing invention and by creating a sort of technological inertia. They might also favor the emergence of novelties by orienting the activity of tradesmen toward a search for quality products. In the sector of clock-making, David S. Landes has observed that guilds "assured the transmission of a level of extraordinary competence," permitting the production of complex and refined instruments, but also that "most guild members were opposed to new things and new ways of making them" and "were compelled to wage a ceaseless battle against the forces of change."[2]

The hereditary transfer of knowledge, the monopolies of production, and the refusal to divulge secrets of manufacture were not limited to the world of artisanship. Many non-regulated trades had identical practices, particularly the ironworks of medieval and early modern times. Hence, in Sweden and Great Britain "a pivotal role was played by a small group of workers among whom close ties provided a framework for the transmission of technological know-how and procedures in the daily conduct of work."[3] In Sweden, in the region of Vallonbruk, "the production of iron by an indirect procedure introduced between 1620 and 1659 by Walloon workers" was organized on a strictly family basis.[4] At the end of the seventeenth century, "the sons of ironsmiths became ironsmiths in their turn."[5] Similarly, in France as late as the nineteenth century, metalworkers employed in the iron industry "refused to instruct strangers." They transmitted their know-how only to those "born as ironsmiths" and jealously guarded their secrets.[6] The hereditary transmission of knowledge from generation to generation was thus a widely spread phenomenon in early modern Europe, which endured into the nineteenth century.

In the early modern period, guild regulations were augmented by those of the state, which were inspired by the doctrines of mercantilism. They were founded on the introduction of privileges and prohibitions to regulate new technology, such as that of seventeenth-century Holland and eighteenth-century England. Under the reign of King Louis XIV, the Minister of Finance Jean-Baptiste Colbert made these practices into a veritable system of government. During the eighteenth century, despite liberal inspiration, they remained common throughout Europe, and in France they became one of the means to introduce the techniques of the British industrial revolution. Such policies sought to favor the development of useful innovations without curtailing established guilds. Yet, rich in contradictions, they often engendered the perverse effect of hindering the emergence and diffusion of inventions.

The regulation of industrial property illustrates the ambiguity of these policies. The text cited from the Venetian Senate, published in 1474, affirmed that if measures were taken to protect inventions from imitation, "we would inspire new talents that will be of the greatest utility to our community." The privileges of inventors became ever more numerous in sixteenth-century Italy and during the seventeenth and eighteenth centuries in European monarchies, because invention was everywhere considered an essential means to achieve the common good, even though its regulation was very diverse. In England the value of the monopoly of invention was a responsibility of the inventor, whereas in France it resulted from the judgment of an academic tribunal. These differences reflected contrary visions of the role that should be played by the state in the development of technology and the economy, but the objective was nevertheless identical: to assure, by the multiplication of inventions, the well-being of the king's subjects and the power of the nation. The survival of the privileges of manufacturing can be explained by the insufficiency of protection provided by patents.

In sum, the culture of secrets and the regulations of guilds or monarchs could hinder innovation, but they did not entirely suffocate it. They did not prevent the methods of workers or artisans from evolving—with some radical transformations being realized in the long term—nor did they prevent the proto-industrial system from developing its innovative capacity. Many activities, especially in rural areas, escaped the grasp of guild regulations or the state. Guild rules slowly weakened in numerous cities during the early modern period, and innovative knowledge began to appear even in the context of trades that were still subject to such regulations. The organization of European society and its system of values allowed a manifestation of forces tending to orient technology toward a search for the amelioration of products and the perfection of techniques. These dynamic forces developed along the following three axes:

1. *The arc of experience.* A system of artisanal knowledge is not static but evolves in a manner such that its effects are never exhausted.
2. *The interaction of trades.* Most productivity requires cooperation among trades, which produces its own dynamic.
3. *The circulation of knowledge.* This has permitted transfers of technology and knowledge among sectors and regions.

The following sections will illustrate each of these aspects of artisanal activity by evoking one or more technological trajectories.

Artisanal Knowledge and the Arc of Experience

In the cluster of empirical knowledge accumulated by artisans, a first level is formed of tacit learning—perceptive in nature, simultaneously visual, olfactory, and manual—which is incorporated in the gestures accomplished in the course of executing the different tasks of a trade. Olfactory and visual know-how play a major role in those sectors using processes of fusion (e.g., metallurgy, glass, or ceramics) in which the choice of superior materials is a decisive element of success. Manual know-how resides in a gesture that can be acquired by the apprentice only through a long process of observation of the movements of a master artisan, followed by personal experimentation. This particular knowledge depends on a profound acquaintance with the instruments of work—be they tools or machines—and their specific procedures of operation. Such knowledge alone is capable of inducing a perfect familiarity with the materials, tools, and machines of a trade, as well as a clear perception of the possibilities that they offer for utilization and transformation. This familiarity is often the result of an extreme division of labor and a no less extreme specialization of skills, particularly in sectors like cloth-manufacturing and clock-making, where, according to Landes, "this hierarchical organization of the shop, based as it was on experience, was the instrument for the transmission of knowledge and skill from one generation to the next."[7]

Artisanal knowledge cannot be reduced to simple perception and manual know-how. According to Bernard Carlson, it forms "a complex of ideas covering the solution of technical or aesthetic problems, the pace of work, the organization of the workshop."[8] Likewise, Paul Feller and Fernand Tourret note that tools are the products of "competing workers' inventions, localized in the personal actions of different distinct individuals" who undergo "a long and patient experimentation" and who possess "an experiential reference of knowledge that surpasses their qualification as laborers."[9] This analysis may refer not only to tools themselves but to all practices of a trade associated with them, which, during many centuries up to the present, have constituted the principal foundation of productive activity. Artisanal knowledge encompassed the learning of rules and empirical practices shared by the entire trade or a particular community of labor. This knowledge was by nature heuristic.[10] It assured the effective development of production by avoiding errors. To these rules and practices, secret or not, forms of complementary tacit knowledge were connected that made their application possible. They were not easily communicated or imitated, and yet they could not be transmitted except by imitation. Like the obsession with secrecy, this was one of the paradoxes of artisanal technology. Manual, perceptive, heuristic, or tacit, artisanal knowledge was the product of accumulated experience and was thus always capable of evolving. Karl Gunnar Persson has contended that the expansion of the medieval economy is not explicable unless one posits a continuous improvement of basic technology achieved by the workers themselves. For him, technological change is a byproduct of the practices of a trade.[11] This analysis applies equally well to the early modern era. The capacity of innovation in the artisanal sphere constituted, in addition to the development of agriculture and the spread of commerce, the primary motor of economic growth in the medieval and early modern eras. It was this capacity that allowed a response to the fluctuation of markets caused by the introduction of new products adapted to the demands of consumers and clients. It has become more effective over the course of centuries, due to a slow but sure improvement in the level of artisans' education, at least in certain regions. Artisans have therefore been able to integrate easily into their trades the new knowledge created by the formalization of empirical knowledge, beginning in the sixteenth century. They have also benefited from the loosening of guild regulations. Artisanal inventiveness affected the totality of groups constituting technology, whether it was a matter of the production of goods for consumption, the process of building structures and hydrological installations, the manufacture of various tools and machines necessary for supplying energy and for motorization, or the treatment and transformation of minerals and metals.

Between the twelfth and fifteenth centuries, a new technological system was created that owed everything to this logic of artisanal apprenticeship. It was developed and modified more or less radically along increasingly diverse trajectories until the end of the eighteenth century. Following is a selection of some artisanal innovations that illustrate the extraordinary expansion of this empirical knowledge. The intention is not to present an exhaustive history of these different sectors of activity but only to describe the cumulative logic of their technological development.

Rural Pottery

The long-neglected history of French medieval pottery was uncovered by numerous archeological excavations undertaken since the outset of the 1980s. Its development followed an arc of cumulative experience. During the twelfth and thirteenth centuries, this technology underwent an initial transformation marked by the appearance of new forms with abundant decorations, by the adoption of new procedures such as oxidization in the process of lead-glazing, and by the construction of ever larger ovens. In the fourteenth and fifteenth centuries, the catalog of potters was considerably extended, and newer forms were adopted that permitted an adaptation to evolving culinary skills and techniques of construction. The production of tin was modeled on that of Italy and Spain. Utilizing the technique of clay columns molded by hand, rather than that of mechanical lathes, potters produced huge vases suitable for storing food or washing cloth. Ovens that were "larger, better constructed, permanent," and sometimes communal were constructed in response to a growing and increasingly exigent demand.¹² Workshops became specialized, even appearing in standardized formats that enabled the development of teams of artisans. Here artisanal innovation assumed a global character. It was no longer concerned only with the amelioration of the manual and visual know-how of workers. Now it also involved the diversification of products and the regulation of the productive process.

Construction

In medieval times, artisanal knowledge was placed in the service of constructing castles, monasteries, and cathedrals, which flourished throughout Europe after the eleventh century. Artisans had to adapt to the use of more diversified and better disseminated materials, such as stone for cathedrals and brick for monasteries. The tools of different trades changed in form and in the method of utilization. Hence, the equipment of the stonemason grew with the appearance of beveled and sharpened hammers and, in the twelfth century, with the reappearance of tools such as the *gradine* (a kind of gouge) and the *polka* (a type of hammer). In fact, as Feller and Tourret explain, in construction and in most other sectors of activity, "the essentials of our technological arsenal were acquired" between the twelfth and sixteenth centuries.[13] This improvement of tools made possible that of other equipment. Since the Roman era, bare walls were built no longer with rough stones but with trimmed rocks joined by mortar. In the Gothic period, larger apparatuses became common. They could be deployed at any height, thanks to progress realized in techniques of lifting. Builders went so far as to polish the non-visible parts of stone blocks in order to distribute their weight better. The organization of construction sites evolved toward rationalization, which had a great influence on the nature of work and skills. Prefabrication was adopted, which explains the uniformity of arches and pillars. Patterns were drawn on parchment and then engraved in stone or sculpted in wood. This method of repeating forms was also applied to framing. The pieces were prepared in advance, then assembled and finished with bolts and mortar. These work methods were closely associated with a growing specialization of trades, which were grouped into workshops whose function was restricted by precise rules. Once again, artisanal know-how evolved toward global innovations that largely exceeded the simple gesture of a tradesman.

Timepieces

In the artisanship of luxury items, the improvement of products found no limit because the spending of princely, aristocratic, or bourgeois clients was so opulent. Clock- and watch-making provide the most obvious model. For example, Parisian timepieces of the fourteenth century attained a mastery of material and form, permitting them to match the sumptuous lifestyle of the Valois court. Among the artistic and technical innovations of the timepieces in use, one may cite the perfect mastery of "enamel covers of embossed objects without recourse to the usual methods used in previous technology such as partitions, hollows, or grooves."[14] This unique procedure, one among many, was not revived in the courts of Europe until the end of the Renaissance.

Cloth-Making

The history of the medieval cloth trade cannot be reduced to the opposition between a classic urban production of high quality, which dominated during the twelfth and thirteenth centuries, and a new rural cloth trade of lesser quality, which developed in the fourteenth and fifteenth centuries. Classical cloth-making originated in the eleventh century and spread thereafter. It depended on an extreme division of labor. As part of the central operation of weaving conducted in the family workshop of a master, a hierarchical structure was created that defined a variety of specialized trades. Some were true artisans; others were workers or female domestic servants.

Important innovations in cloth-making were introduced beginning in the twelfth century. The main one was the replacement of traditional hand-weaving (usually limited to domestic production) by the horizontal chain-weaving of two persons. The felting mill, attributed first to an Italian monastery in AD 960 and later to Normandy in 1086, had only a limited diffusion before the thirteenth century. Foot-pedaling, which guaranteed a better quality of cloth, survived for a long time. It took a while for mechanical felting to be improved to the point that it could rival the foot-driven process. These operations lent to a beautiful cloth its appearance of uniform texture with even coloring and invisible strands of thread. Such techniques created the prosperity of Flemish drapery after the twelfth century. The quality of products was assured by a very precise definition of their characteristics and a strict control of operations. This system depended entirely on the know-how of workers and artisans: it was they alone who guaranteed quality and ensured the reputation of one textile center or another. In such a system, the creation of new methods could be achieved only within very restricted limits.

This system was disrupted in the course of the twelfth and thirteenth centuries by three innovations: the *arçon* (a means of dressing or carding wool that was invented in 1227 at Toulouse), the metal crochet hook, and the spinning wheel. These three novelties enabled the use of short fibers furnished by wool-combing and also by merino sheep. They allowed a mixture of colors and the simultaneous stretching and twisting of thread, which led to significant increases in productivity. This technology initially stimulated an improvement of classical draperies, but it also led to the production of lighter tissues. Threads for weaving were carefully carded and then spun with the aid of a spinning wheel. In the fourteenth and fifteenth centuries—at first in Spain, where raising merino sheep was developed, then throughout Europe—the Vervins cloth appeared. Made from carded wool and of excellent quality, it was moreover cheaper than classic combed wool due to being manufactured through a combination of the three innovations cited above. This cloth had a future.

In this fashion, a complex technological system came into being. It depended on the transmission of know-how from generation to generation and scarcely evolved further until the early nineteenth century, when machines appeared that were capable of performing the same operations as workers, female domestic servants, and artisans in workshops. The organization of labor founded on a combination of complex accumulated knowledge, which was essentially tacit, empirical, and closely coordinated, was thereby perpetuated until this period. Gérard Gayot has precisely described "the twenty changes of handicraft that occurred (at Sedan) in the course of twenty successive operations" that transformed the textile industry.[15] His long discussion reveals the precision with which individual gestures and their elaboration were regulated. The mechanization that followed in the industrial revolution took its place in the logic of mastering this process. During the eighteenth century, these methods of labor became more refined. Whether in manufacturing paper or glass, they still depended on a succession of gestures and specific operations or, as in the production of pins described by Adam Smith, on the decomposition of gestures into simplified sequences. An evident continuity of process was thus established between the modes of production in the early modern period and those of the industrial revolution.

Watermills and Windmills

Those who invented mills were incontestably the heroes of a technological adventure that developed more and more complex machines, integrating the know-how of previous times into mechanical conceptions of the industrial revolution. The constructors of mills, experts in the technology of cogwheels and the transmission of power, were able to extend their skills into other applications after the closing centuries of the Middle Ages. The functioning of mills presented innumerable difficulties, evidenced by the incessant repairs needed to maintain them. It was necessary to adapt their mechanisms and materials to constantly increasing activities. Contrary to conventional wisdom, the technology of mills was neither simple nor stagnant. Their conception gradually created a body of knowledge about hydrological power, and their construction required specific know-how among carpenters, cabinetmakers, masons, wheelwrights, and blacksmiths. The invention of watermills and windmills demanded dual skills of conception and production: the conception had to take into account the unique characteristics of the source of energy utilized, while the production involved a close acquaintance with the materials used, as well as the means to treat and assemble them. For a long time, the principal material employed was wood, which justified the prominence of carpenters and cabinetmakers in the hierarchy of tasks. But the use of metal gradually ensued, giving a more important role to blacksmiths and other metal tradesmen in the course of mechanization.

Until the seventeenth century, steam and hydrological power played a rather minor role compared to animal and human energy. For a lengthy period, the use of watermills was confined to the grinding of grain. It was not until the thirteenth century and still more the second half of the fifteenth century that this technology truly unfolded, thanks to progress in the transmission of power, which led to the innovation of various new types of mills.

Three types of fixed watermills coexisted in Europe: mills with horizontal wheels, common in southern mountainous regions; mills with submerged vertical wheels, adapted to the rapid and irregular flow of rivers; and mills with water-level wheels, appropriate for the constant flow of calmer rivers. Each of these technologies was the product of adaptation to local conditions. The mill with a horizontal wheel was widely dispersed in Italy, southern France, and Catalonia. Its use was essentially limited to milling grain and to sawmills. Frequently used in isolated villages, it was easy to construct, but it had no gears and could not be accelerated, unlike mills with vertical wheels. The submerged wheel was simpler than the water-level wheel, but its output was also feebler. Vertical wheels could be equipped with cups or buckets whose form and function were the object of numerous experiments and improvements, starting in the seventeenth century. Their performance became increasingly regular.

A common knowledge of these different types of wheel began to spread, establishing the basis of hydrodynamics, which depended on the transfer and confrontation of technology among regions. Thus, in the fourteenth century, a system coming from France that combined the horizontal wheel and procedures of the submerged waterwheel appeared in Tuscany. As a result of these empirical experiments, between the sixteenth and eighteenth centuries, builders succeeded in increasing the annual period for the use of mills and the power of their installations, although this did not exceed 10 horsepower before the eighteenth century.

It was in Flanders and Holland that the technology of windmills, which quickly advanced in the thirteenth century, attained its highest degree of sophistication. The combination of parts and mechanisms had to be adapted to local uses and to working conditions that were often extreme. This effort was concentrated especially on mobile units, wooden shafts, flywheels, other wheels, grindstones, and the like. Hence, a science of mills was created, and despite being very diverse from one region to another, it was subjected to some general regulations. From the sixteenth century on, knowledge "in the slightest detail" was required of a great number of mechanics of transmitting and transforming motor force as well as the application of hydrological energy and the function of grindstones and hammers.[16] Contrary to the usual assumptions, the use of metallic parts does not date from the eighteenth century but was common since the fifteenth century. In Flanders, blacksmiths deployed a multitude of welding procedures to reinforce the solidity of metal items, and the production of these parts was reserved for blacksmiths specialized in the manufacture of iron in mills.[17] In the sixteenth century, some important progress for windmills used to dry land was realized in Holland, and the Flemish and Dutch technology spread throughout Europe. It was thus that England, which until the eighteenth century had imported the machinery of windmills and watermills from the Continent, began to produce its own mechanisms. A special class of mechanics, called millwrights, appeared; they were assigned to the production not only of mills but also of pumps and all sorts of other machines. They adapted their know-how to the burgeoning use of iron in the production of mill parts, and they naturally became experts in mechanics and metalworking for the manufacture of increasingly complex machinery, such as textile machines and lathes. In this way, they played a significant part in the emergence of the English and European mechanical industry during the industrial revolution.

Other Trades

Several other trades besides those concerning mills participated in the formation of know-how necessary for the rise of mechanical industries. The history of regions that witnessed the birth of the English industrial revolution, such as Lancaster and Yorkshire, illustrates the diversity of accumulated know-how indispensable for its expansion. It was from the beginning of the seventeenth century that artisans and domestic workers became specialized in the treatment of metals in the production of mechanical objects. Blacksmiths, smelters, fitters, hammerers, and manufacturers of all sorts of tools formed the first body of know-how and the initial professional cluster necessary for the fabrication of textile machines and steam engines in the eighteenth century. By utilizing this diverse knowledge, artisans from the regions of Birmingham, Sheffield, Manchester, and Nottingham began to conceive, at the beginning of the seventeenth century, a vast number of new consumer products destined for households and dining tables or for personal ornamentation, such as buckles, broaches, and decorative buttons of all kinds, made of polished steel alloys or imitation gold. The procedures were expensive and required fastidious labor, since the ornaments had to be assembled or engraved by hand. The tools, presses, buffers, and lathes were improved as a result of "an infinity of tiny improvements of which each artisan jealously guarded the secret," according to one foreign writer.[18] In reality, secrets could not be kept for long in such a milieu; hence, it was possible to derive a large number of new products. These innovations made the reputation of Birmingham, which in the second half of the eighteenth century became a capital city of artisanal invention. An observer commented in 1777 "that here the inventor who creates the model of a machine ... may quickly gain proof of the good quality or the imperfection of his conception, thus avoiding the difficulties, expenses, and delays that its testing would require." Birmingham thereby acquired "a community of inventors stimulating each other to make the best products or utilize the best procedures."[19] The demand for materials created by the expansion of this activity, along with the ensuing need for quality metals furnished by the iron industry, provided a major impetus for the English industrial revolution.

The Interaction of Trades

Cooperation among the different trades in the process of technological change played a dual role: the fusion of know-how from diverse fields and its dissemination. These were necessary conditions for the appearance of innovations in the system of empirical knowledge, since objects or technological procedures were solely the product of a successful synthesis of more than one know-how. Cities were the preferred site of interaction among the different trades, an interaction rendered all the more indispensable because guild regulations pushed the division of labor to its extreme limit. The use by one trade of innovative knowledge drawn from another was a particular form of transfer that produced an identical result—namely, integrated learning. Many technological trajectories of the medieval and early modern periods illustrate this dynamic force of exchanges among sectors.

Printing

Printing was the collective invention par excellence. It was perfected during the years 1435–1450 in the milieu of metal arts. Since the fourteenth century, Europe had been experiencing an intense rivalry in the world of literature. The force of this current was all the greater in Germany because the intellectuals of that prosperous country felt a sense of inferiority in this sphere. Furthermore, a desire for elaboration and exchange was everywhere present. It was in this atmosphere of collective rivalry that printing appeared, but it was the availability of cheap paper that made possible its progress. The production of paper, introduced in Spain in the middle of the twelfth century, soon became mechanized in Italy and spread throughout Europe in reaction to a scarcity of parchment. Others besides Johann Gutenberg had noticed the potential offered by "mechanical writing," but the significance of his contribution cannot be disputed. He was the only person to achieve a synthesis among technologies belonging to different sectors. Using his metallurgical know-how and his knowledge of jewelry-making, he perfected an alloy to produce movable type that was both easy to cast and very hard when set. The letters were formed in a copper mold and outlined with the aid of a stiletto punching device used in making jewelry. Moreover, Gutenberg invented a manual smelting pot that enabled him to vary rapidly the size of letters and to control their form with great precision. Finally, he borrowed from cloth-printing and wine-making the technology of pressing. Such a synthesis could be achieved only in a milieu like that of Mainz, where there were several adjacent crafts. Gutenberg knew how to coordinate them because he was located at the crossroads of these various trades.

Clock-Making

The construction and maintenance of mechanical clocks in the fourteenth century constituted "a complex and diverse art marked from the beginning by the collaboration of several trades and by the division of labor." Thus, "every machine was the collective product of a team" that united various skills.[20] Groups of specialists selected for their mastery of mechanics were gradually collected into workshops. Those first manufacturers were most often blacksmiths and cobblers, or else coppersmiths and locksmiths. The cooperation among such trade groups was further emphasized with the invention of the wristwatch in the sixteenth century, since it was as much an ornament as a mechanical device. At the heart of the watchmaker's trade, the different operations began to fragment, as was the case with cloth-making and metallurgy. It was notably the appearance of springs in the fifteenth century that obliged clockmakers to call upon those capable of producing and mastering metals, all the more so since "no true artist was prepared to reveal which product he used to dip his hot metal."[21] This diversification of skills tended to diminish during the seventeenth century, but master clockmakers continued to rely on outside specialists to perform some of their operations.

Scientific Instruments

The history of scientific instruments illustrates both the necessity of cooperation among trades and the negative effects of their division and of excessive regulation. Advances in the production of scientific instruments in the sixteenth century paralleled the progress of closely related mathematical and astronomical instruments, especially in Paris. Their development required a capacity for theoretical and manual skills. The number of manufacturers rose in the seventeenth century because of growing demand, and the trade was organized through patents in 1608. During the second half of that century, production was extended to new devices: telescopes, microscopes, barometers, thermometers, air pumps, machines. An identical evolution occurred in Holland and Italy, where the scholarly demand was particularly important. It was a Dutch artisan optician who invented the magnifying glass, which Galileo transformed into a telescope, a pure product of artisanal know-how. Galileo and many other scholars thus contributed to the inception of the first instruments of scientific observation. To that end, it was essential to call upon a great number of trades, such as manufacturers of mirrors, stone polishers, and producers of eyeglasses, which were very different from one another in this know-how. In Italy, the very first astronomical lenses were produced with the primitive equipment of mirror factories in Murano and polishers in Florence. The results were mediocre, so other trades were sought; these included cardboard-makers and bookbinders. Finally, it was the transfer of know-how from printers and smelters of brass and copper that ensured the ability of turners and glassworkers to satisfy demand. The centers of production, while remaining numerous in Italy, migrated to the Netherlands and southern Germany, then to France and England.[22]

The French model gradually weakened, however, and in the second half of the seventeenth century, French production underwent a notable decline. Guild regulations hindered producers in very disparate trades from cooperating, launching new products, utilizing new materials, or increasing the size of their workshops. They were consequently incapable of adapting to the evolution of a market in full expansion. In Paris, the organization of luxury trades was too rigid to permit French producers to integrate changes required by the demand for high-precision instruments. Meanwhile, English industry, whose origins dated back to the seventeenth century, developed vigorously. As a result of greater freedom, its large rationalized workshops manufactured series of increasingly diverse products that better served the exigencies of precision. The type of trajectory that characterized this sector was identical to that of clock-making and industrial mechanics, from which it partially emerged, and it was to become the master discipline of the nineteenth century.

Furniture

The Parisian furniture industry did not cease to innovate in the seventeenth century and attained its highest degree of prosperity in the eighteenth. The perfection of its products resulted from a continuous deepening of know-how, which, far from bogging down in routine, led to a multiplicity of remarkable innovations, both technical and aesthetic. This success derived in part from a clear perception of the products that would be favored by high-income consumers. It also stemmed from a dual complementarity between designers and artisans, on the one hand, and among different artisans, on the other. The extraordinary sophistication of decoration would not have been possible if artisans had not possessed a graphic model that was innovative and perfectly designed, which is why one cannot separate the accomplishments of cabinetmaking from the perfection attained by French decorative design in the eighteenth century. This complementarity was identical to that associated with industrial design, connected to descriptive geometry, in the emergence of the mechanical industry. As Jean Meuvret has remarked, "the production of deluxe furniture in the eighteenth century required the collaboration of numerous trades with clearly defined attributes." Furniture became "the collective work of Parisian artisans, even though woodworkers remained at the center of all such enterprise."[23] They called upon carvers and decorators. Cabinetmakers enhanced drawers with silk and desks with leather. Ornamentation was provided by wood sculptors and shaping by lathe operators. Seats were turned over to wood decorators or painters, upholsterers, and lace-makers. The organization of artisanal communities was not as rigid as was previously thought. The immigration of foreign workers, especially Germans, and the sheer number of activities created a system of free competition. In sum, artisanal communities played a beneficial role. They assured that certain norms of production were respected and maintained a degree of control over quality. Above all, they made possible effective cooperation among the trades by imposing rules and limiting conflicts. They afforded these relationships a stable and well-defined framework that allowed for the progress of practice and knowledge, thanks to daily exchanges established among trades in order to achieve the perfection of their products—the fruit of a collective effort.

The Uses of Coal

In the history of the industrial utilization of coke, such exchanges and complementarities appeared more widespread but were no less significant. Derived from mined coal, coke was substituted for wooden coals in large ovens and forges. This discovery, dating back to 1709, has been attributed to a master blacksmith, Abraham Darby, whose innovation required a full mastery of carbonization and thermal energy. It was the last step in a long process of gradual, collective improvements, stretching back to the fifteenth century, which permitted this substitution for the greatest possible number of applications. The development of the industrial use of coal at the end of the Middle Ages is commonly associated with the scarcity of wood, which became progressively more pronounced, especially in Great Britain. The British very rapidly grasped the fact that the only effective solution was the substitution of mined coal for wood, but this involved serious risks due to the composition of coal. The operation was susceptible to the deterioration not only of materials that came into contact with coal but also of the installations (furnaces, forges, melting pots, and other containers) in which the procedures were executed. That is why the introduction of coal extended over a long period in most industries and was accomplished gradually due to a succession of incremental innovations. At first, the sectors of application were those where coal did not come into contact with the material being treated. The baker, the dyer, the brewer, and the producers of salt or lime used coal very early on, although it was necessary for those industries to overcome prejudices associated with its use. Coal was accused of lending a bad odor to bread and beer and of weakening bricks and glass. To respond to these accusations, brewers utilized a purified and desulfurized form of coke. In the glass industry, the control of pollution was improved after 1670 with the appearance of leaded flint glass.

But it was the progress realized in the treatment of non-ferrous products that opened the way to coal-driven metallurgy. In the course of the seventeenth century, techniques allowing the fusion of non-ferrous materials were elaborated with the aid of coke. The English thus succeeded by 1680 in constructing reverberatory furnaces and more efficient chimneys. These furnaces provided a key to the treatment of lead, copper, tin, and, around 1700, brass. The English attempted during the sixteenth century to ameliorate the materials used in melting pots by inventing fireproof products specific to each of the metals treated. They perfected the technology of fire by controlling better the loading and distribution of coal in the furnace and by adapting the rhythm and operation to the nature of combustibles. Darby's work fell into the flow of these new practices. Before renting a large abandoned oven in 1708, he had managed a brass foundry and was thus able to transfer his know-how from one sector to another. Darby's new method did not spread elsewhere until after 1740 because for a long time he kept secret the procedure that joined the use of coal to a course of sand, which made the sand track much more fluid and afforded a maximum of efficiency to the use of coal.

Accordingly, writes John Harris, this innovation was the fruit of a combination of "common sense, imagination, and the method of trial and error, which could be applied without educational or institutional inhibitions." The empirical method in this case did not encourage routine but favored the evolution of different trades. Technological knowledge was refined by "qualified experienced workers and not by entrepreneurs."[24] It was, one could say, appropriated by workers and transmitted by means other than writing. The artisans and workers who achieved a synthesis of this know-how listened to other trades, interpreted their messages, and found the means of a genuine convergence among promising forms of knowledge. Their discovery was the fruit of collective innovation, which was gradually developed in the experiments of construction sites and workshops.

The Circulation of Knowledge

The circulation of knowledge from one generation to another, and also from one trade or region to the next, was one of the major sources of innovation. It provided an essential impulse in the construction of fresh knowledge and made possible the confrontation and then fusion among different practices of the more or less extended network of designers and operators.

The intergenerational transmission of knowledge appropriated by artisans and workers was accomplished by speech, the repetition of gestures, or graphic designs. These modes of transmission were practiced on the worksites of Gothic cathedrals in the thirteenth century. Blueprints were used there to implement construction but also to instruct young apprentices. Spread out on a stone or platform, they served to conduct demonstrations. This instruction found a place in the process of apprenticeship that lasted over time and assured the perpetuation of knowledge.

Building Sites

The great sites of construction or equipment, like those at mines, dams, or towns, grew exponentially in Europe after the twelfth century. They were ideal locations for the intersectorial and interregional transmission of knowledge because they brought together workers and qualified artisans from various areas and different trades. Talents and skills drawn from all sides were assembled, and the transfer of knowledge was not greatly hindered by guild regulations. The circulation of learning related to these sites was thus a major avenue for the diffusion of innovations from one region to another and from one sector to another. Construction sites made possible the confrontation and combination of various techniques in the light of experience. This analysis applies perfectly to several large religious and civic sites that have been well-documented, such as the cathedral of Amiens in the thirteenth century, the Milan cathedral in the fifteenth century, the mines of Jacques Coeur at Pampailly toward the end of the fifteenth century, and the mechanical works of Marly in the seventeenth century.

These large sites illustrate the role played by the permanent or temporary migration of trained artisans in the circulation of knowledge and the transfer of technology. Intense already in classical antiquity, these migrations never slowed and increased rapidly after the eleventh century. The travels of artisans in medieval and modern times, like those of engineers, became essential factors in the expansion and also the unification of technological learning. During the entire early modern period, they remained one of the principal channels of industrial espionage and one of the favorite means of technological transfers engaged in by mercantilist states as part of their competitive economic struggles. The mobility of professors and students, which had an established tradition, also intensified in early modern times. A veritable European market of knowledge was thereby formed and opened to foreign lands.

The history of technology is rich with examples illustrating the mobility of artisans and experts. The Venetian republic "is distinguished by a long tradition of selectively welcoming foreigners, Byzantine mosaic workers, silk experts, Germans who made daily bread for Venetians in the fifteenth century and who doubtless brought the secrets of refining silver before printing."[25] Since the thirteenth century, the glassworkers of Altare in Italy departed every year in seasonal migrations, to Italian workshops at first and then to France during the sixteenth century. These shops "paid a financial sum to the guilds of Altare to obtain workers."[26] Bernard Perrot, a native of Altare, took up residence in France in 1647 to found a glassworks at Nevers. He was the actual inventor of the glass casting used by Saint-Gobain to make large mirrors. Another example is medieval rural pottery. About 1380 Spanish artisans working for the Duke of Berry introduced tin-making into that region. In England, there were some 500 Norman workers who, under the patronage of the Tudor state, introduced the production of iron between 1490 and 1540. In Sweden, Walloons, who settled with their families, perpetuated the same process between 1620 and 1650. Two Armenian artisans were brought to France in 1672 by Marseille merchants so that they could reveal to French workers the secrets of Indian production with Hindu methods, a process that spread rapidly despite many restrictions. This was similar to the migration of French Huguenots to England, Germany, and Scandinavia after the revocation of the Edict of Nantes, an essential factor in the rise of several major industrial sectors in those areas. The frequent migration of artisans between France and England in the sectors of glassware and metallurgy in the eighteenth century became a mutual source of technological transfer, although their social integration always remained difficult.

Voyages by engineers in the eighteenth century were a significant element of their education in schools created for that purpose and were a principal factor in many of their careers. French mining engineers of that time were directly exposed to German technology, as were Continental metal or mechanical engineers at the beginning of the nineteenth century to English techniques, as well as European engineers in the late nineteenth century to American technology. Hence, in the early modern era and beyond, a vast global market of technological knowledge was created.

Chapter 2

From Artisan to Expert

Ever since medieval times, formalized and sometimes even codified learning was grafted onto artisanal knowledge. It was either the distant heir of practices in antiquity or the fruit of "discursive" transcription. This transfer was achieved by artisans themselves or by external observations of artisanal practice. Thus, emerging from within artisanal circles or without, expert knowledge was created that would soon be identified with engineers in industry and large enterprises or with architects. Their competence was initially derived from the appropriation and reinterpretation of tacit and empirical knowledge, based on their study of artisanal skills and labor practices, and was enriched by personal experience acquired through the exercise of their profession or by the results of experimentation. Their mission was to devise fitting responses to breakdowns and malfunctions and then to formalize the procedures in use or to conceive new ones. This tendency explains why, beginning in the fifteenth century, invention was more and more clearly viewed as the result of individual actions. The appearance of legislation on patents and the support afforded by mercantilist states to inventors illustrate this evolution.

In each sector, an "engineering science" was thereby composed of formalized and codified knowledge that could be transmitted in writing. This development did not cause the disappearance of tacit knowledge or empirical techniques; on the contrary, in all sectors the concrete application of principles and of "scripts" defining procedures could be effective only through what some authors have called "tacit residues." The practices and techniques gathered from such innovations encouraged new tacit knowledge and created new skills, while others were condemned to disappear.

The Appropriation of Artisanal Knowledge

This evolution depended on a slow but irrepressible tendency toward the formalization and codification of empirical knowledge, founded on more or less successful attempts to theorize and made possible through the transmission of knowledge by oral, graphic, or written means. Hence, a transformation occurred in the midst of the artisanal world. Certain guild regulations that precisely described procedures of production might have favored the emergence of this formalization. But in particular, the artisan, who possessed a direct understanding of skills, was able to acquire a competence that enabled him to analyze techniques and formulate them with sufficient coherence to provide a formalized expression. That was the case of a German technician who equipped the mines of Jacques Coeur at Pampailly or the Liège engineer who conceived the machinery of Marby. They were not only artisans but experts whose competence was internationally recognized. Such was also the case of those who constructed windmills or watermills in all regions of Europe from the thirteenth century on.

In many trades, a redistribution of skills began at the end of the Middle Ages and gained momentum especially in the sixteenth century. Masters became, as Bert de Munk has said, "grand master entrepreneurs," like the cabinetmakers of Antwerp. With advanced skills that set them apart from manual workers, these grand masters became true managers, designers of projects, and inventors.[1] By facilitating a clear formulation of basic elements in empirical knowledge, literacy accelerated the spontaneous process of appropriating knowledge within the artisanal world. A large number of artisans were thus transformed into experts, precursors both of professional engineers and industrial entrepreneurs, just as medieval masters of construction had prefigured the modern architect.

Another path to formalization was that of an external view cast upon artisanal practices by different actors—intellectuals, architects, entrepreneurs, master artisans, or others—who obtained expertise in a particular realm of knowledge that could be defined as "an acquired combination of tacit and formal learning in specific contexts."[2] Since the end of the Middle Ages, they have often been called engineers, but the designation of "experts" is preferable in order to distinguish them from professional engineers, who, after the seventeenth century, were granted a precise social status corresponding to that title.

Whether their expertise was purely artisanal or derived from elsewhere, experts of all sorts who circulated throughout Europe between the twelfth and seventeenth centuries possessed knowledge that was no longer basic knowhow. Their skill could be applied not only to this or that particular operation but to an ensemble of procedures capable of being coordinated. Their knowledge also differed from that of a simple artisan by integrating formalized knowledge that was of a mathematical and geometric nature or was descriptive and discursive. That was the case for master builders of Gothic cathedrals and also master Antwerp carpenters of the sixteenth century, who were capable of employing geometric calculations in order to represent much more complex forms than those imagined by apprentices or itinerant laborers. Architects, entrepreneurs, and engineers were plunged into the concrete practice of construction and production through the utilization of tools offered by mathematics, drafting, and geometry. In the eighteenth century, the builders of mills became specialists in mechanics and hence true experts. British millwrights, armed with the same knowledge, played a central role in the process of mechanization leading to the industrial revolution.

To develop their skills, experts did not confine themselves to initiating mathematics and geometry. They also applied strategies of appropriation, observation, and transcription in a plain written language. In such cases, the process of formalization involved three steps: observation, description, and the definition of rules based on those observations. In his *Discours préliminaire de l'Encyclopédie*, d'Alembert described the method he used to communicate artisanal knowledge to the reader. "There are," he wrote, "trades so singular and maneuvers so subtle that, apart from doing the work oneself, from operating a machine with one's own hands, and from seeing a product being created with one's own eyes, it is difficult to speak of them with precision."[3] His intention was, by conversing with artisans, to transform their knowledge into learning that was transferable in words and especially in drawings, not only of gestures performed during operations, but also of the machines themselves. It was an attempt at formalization that relied on the transmission in writing or drawing of knowledge contained in labor practices and on the transformation of tacit learning into codified learning. Innovations in the use of coal for the treatment of non-ferrous objects, which as noted appeared in Great Britain in the seventeenth century, were not the object in that country of any description in technical publications. Nevertheless, they gained the attention of foreign ironmongers—French and Swedish in particular—because they recognized the radical character of these innovations and the importance of transferring them to the Continent. This type of appropriation became one of the essential aspects of industrial espionage, which was practiced by all the mercantilist states of the early modern period. The education of French mining engineers in the eighteenth century included obligatory trips abroad, in the course of which they were expected to discover and appropriate techniques of the countries visited.

The Role of Writing

The effects of printing should be located within this general framework. Ostensibly, it merely extended previous efforts to spread the learning of certain experts by writing. But in reality, it gave to this diffusion a new dimension and significance, permitting a radical transformation of technological language and its penetration into artisanal and industrial circles. Moreover, it highlighted the polemics and confrontations concerning different existing procedures. After the fourteenth century, established specialists in this or that sector of activity, such as mining, metallurgy, artillery, fortifications, mechanics, or agronomy, as well as those with knowledge based on experience and concrete observation, appeared throughout Europe. They became advisers to princes, cities, or important individuals, and they were also authors of treatises describing hydrological machinery, other mechanical devices, or instruments of war. Of course, these types of treatises were not entirely novel, since they had existed since the twelfth century. What was new was their content, which was often oriented toward a critical analysis of procedures as much as a simple description of them. These publications, which opened the way to fresh solutions and inspired the notebooks and drawings of Leonardo da Vinci, had a wide distribution in the sixteenth century thanks to printing. Among the works printed at that time, one must cite the *Pirotechnia* of Vannoccio Biringuccio. Published in 1541, it treated those industries that used fire in their production, ranging from distillation to metallurgy and including cannon powder. The *De re metallica* of Georg Bauer (named Agricola), published in 1556, studied occupational diseases in the mines and, in the process, created a synthesis between the mining and metallurgy of that era. As a result of these and similar treatises, disciplines such as chemistry, metallurgy, and agronomy evolved toward a more concrete approach to phenomena based on the rational observation of artisanal practices. Hence, after the sixteenth century, existing technological knowledge was overturned: knowledge that had been restricted to small groups was transformed into knowledge that was now widely shared and articulated in formal principles. This enabled the massive transfer of learning from one region or sector to another. One of the most striking aspects of this proliferation was the spread in the eighteenth century of encyclopedias and dictionaries containing analyses of arts and trades. The *Encyclopédie* of Diderot and d'Alembert thus devoted forty-four pages to the production of glass, thirty-three to masonry, and twenty-five to mills.

The slow transformation of technological learning was closely associated with the appearance of new actors in the construction of knowledge. Besides artisans who became recognized experts in their trade, from within or outside of such a milieu specialists emerged who possessed useful knowledge and soon became engineers. They declared their intention to dominate not only processes of production but the totality of a discipline. The science of engineering, autonomous from the "natural philosophy" of savants, thus expanded inside of workshops and without. An examination of French patents in the second half of the eighteenth century reveals among inventors—whether artisans, engineers, or manufacturers—a growing mastery of technological culture. European and American industry of the nineteenth century would be the product of this engineering science.

Knowledge of Experts

Examining the trajectory of several sectors should illustrate the nature of expert knowledge, which appeared very early in some areas such as construction, hydrology, mines, warfare, and mechanics. Master craftsmen and military commanders had recourse to the skill of artisans who were also themselves true experts, insofar as they not only undertook the performance of a particular task but also understood a cluster of separate albeit complementary operations.

Masters of Medieval Construction

On the building sites of the medieval period, the means of describing, teaching, and transmitting to apprentices the projects of master builders and the plans for labor organization ceaselessly evolved. Beyond the spoken word, blueprints were the principal agent of this transformation. These drawings could be executed only by workers who had acquired some knowledge of geometry. The documents created by workers possessing such knowledge and those who did not displayed a marked difference, thereby establishing the pre-eminence of master builders. Having received "an education of high quality," they became real architects "whose activity was concentrated on blueprints and drawings."[4] This education was often acquired through heredity within veritable dynasties, for example, as the Montreuil family or the Chambiges in France, who were organizers of building sites and trades in the thirteenth and fifteenth centuries. Thus, beginning in the thirteenth century, a distinction became evident between a master builder qua architect and the foreman of a construction site. The former was distinguished by his attention to the precision of measurements and models, as well as the mechanics and tools utilized. Master builders introduced innovations in the conception of edifices and contributed to the invention of a new architectural language that permitted unprecedented technological and aesthetic performance. These accomplishments were the result of close cooperation between artisans, workers, and master builders.

A similar evolution occurred in other sectors, such as large hydrological installations and mines or trades that involved the military arts. Experts were gradually formed into identifiable groups in those sectors as well. Detached from their artisanal bases, they were sometimes even integrated into monarchical administrations, which played an increasing role in technological trajectories. In the realm of hydrology and mines, such expertise derived from long experience in the exploitation of these systems. In fortifications, it depended on a rational analysis of the evolution of offensive weapons. In mechanics, it was the product of careful observation of breakdowns, associated with a clear understanding, gained through practice, of the possibilities offered by different materials and processes.

Hydrology and Mining

The know-how about hydrological power, amassed in Burgundy during the eleventh century by the monks of several Cistercian monasteries, spread throughout Europe. This was due to experts sent by that order to assist in the creation of new establishments, whose management was then imitated by towns. The large medieval hydrological installations were conceived and constructed by artisan plumbers whose knowledge had been passed down for generations from antiquity to the late Middle Ages and who perfected and extended the proliferation of networks. Furthermore, new skills in mechanics, necessary for the creation and maintenance of windmills and watermills, completed this cluster of knowledge. Coherent new systems were thus inaugurated, and a European network of hydrological technicians, who became increasingly active after the fifteenth century, was formed. These developments could be observed in the Netherlands, where the drying of land to create more agricultural territory was undertaken in the twelfth century. It reached its limits at the end of the fifteenth century because of difficult climatic conditions and the unduly intensive use of peat, which weakened the dykes. Before 1500, institutions were improvised to master and administer drainage systems. In the course of the sixteenth century, following the replacement of the corvée by taxes, the maintenance of the network was entrusted to a corps of trained specialists. Several major innovations were introduced after 1550, particularly in the technology of windmills, which sparked a veritable explosion in reclaiming land. Likewise, the technician Rennequin Sualem, who built a machine to irrigate the gardens of Versailles under Louis XIV, was the descendent of a family of watermill experts who had acquired their reputation and expertise in the region of Liège. They designed equipment to pump water out of mines, then to service city fountains, and finally to nourish the gardens of elevated castles. Sualem installed canals of cast iron 150 meters long at the castle of Modave to raise water from the River Hoyoux.

During the fourteenth and fifteenth centuries, new mining techniques appeared, spread for the most part by German experts and workers. As a result, the use of underground transversal corridors was notably expanded in all of Europe. This technology, placed into service in the mines of Jacques Coeur at Pampailly in France, permitted the utilization of little carts on wheels, called "mine dogs," thus allowing for more ventilation shafts and the possibility of connecting the strata of different levels. Therewith developed the use of waterwheels to grind minerals and to operate smelting ovens in the production of silver, as occurred in Germany around 1450. All of this knowledge was bruited with the aid of writings and drawings, particularly those in Bauer's *De re metallica* of the sixteenth century.

Military Engineers

Within the military sphere, texts reveal the existence of "master engineers" who specialized in the production and maintenance of catapults and other launchers after the end of the thirteenth century.[5] Many Renaissance engineers invented ballistic devices and specialized in the art of siege or defense, thereby creating a true science of fortifications that was revolutionized after 1450 by gunpowder, as was effectively demonstrated by the Italian wars. In the sixteenth century, Italian military architects like Antonio Sangallo and Michele Sanmicheli became the undisputed masters of this discipline, which was based on a mathematical approach to the problem of ballistics and on a more empirical understanding of durable materials. These experts served monarchies throughout Europe, such as in France where "royal engineers" who constructed fortifications were, in the great majority, Italians.

These eminent specialists in architecture and in hydrological, military, or mining technology did not form a true professional community. They remained isolated individuals, often belonging to family dynasties that specialized in the knowledge of one or another of these technologies. Their social recognition and power rested entirely on their personal reputation and not on that of an established corps, as would later be the case for fortification engineers, English military engineers, German mining engineers, and the French engineers of Ponts et Chaussées.

Metalworkers and Mechanics

Mechanical tools were introduced in various branches of production, such as textiles and metallurgy, during the twelfth century, but it was not until the closing centuries of the Middle Ages that machines appeared for grinding grain, making textile pulp, crushing minerals or olives, and forming products that included wire drawers and flattening mills in the treatment of lead, copper, or gold. Machines also served to operate devices as diverse as mechanical clocks, mathematical instruments, construction and ballistic machinery, and mechanical toys. The most broadly distributed of these objects were windmills and watermills, whose widespread adoption was made possible as a result of a radical transformation of medieval mechanics.

The ensuing evolution tended toward the formation of artisanal expertise beyond the strict limits of a particular trade. In the fourteenth and fifteenth centuries, a genuine applied mechanics developed that became, after architecture, a leading technology, attracting others in its wake. Its development was contemporaneous with the appearance at universities of scientific inquiry in the form of many treatises involving the description of all sorts of machines. If these writings, as suggested, directly inspired the technical drawings of Leonardo da Vinci, his mechanical genius was above all due to his perfect understanding of the practices and know-how of Florentine artisans.

In sum, one can say that technological expertise was at the same time a product of artisanal knowledge and its culmination, because it opened the way to the formalization of knowledge. Socially, it facilitated the appearance of recognized specialists in several disciplines and of new actors in technological change who foreshadowed professional engineers and entrepreneurs. Intellectually, it paved the way for a new form of knowledge that was founded, partly as least, on drawings and writings. It tended to define the norms of production with general rules of operation and to develop long trajectories that were not only cumulative and empirical but also logical and deductive.

Chapter 3

Formalized Knowledge

Between the middle of the seventeenth century and the outset of the nineteenth century, the influence of experts on the development of knowledge was progressively strengthened. It took hold in virtually all realms of technology as a corpus of formalized learning was diffused in countless treatises and technical encyclopedias, and it was taught in universities and schools. Thus, the expert was transformed into a professional engineer who enjoyed recognized social status and prestige. The merchant-manufacturer and eventually the engineer became entrepreneurs capable of organizing the knowledge and skills of several crafts in order to produce and sell goods. They were also capable of conceiving or adopting innovations to improve the quality or reduce the costs of production. Many of these experts, engineers, and entrepreneurs were increasingly near to the world of intellectuals. They were inspired by methods of experimentation and deduction that appeared with the "scientific revolution" of the seventeenth century and were thus influenced in their actions by the evolution of scientific thought.

The Professional Engineer

Recognition of the competence of experts was strictly individual; recognition of that of professional engineers was collective. From the seventeenth century onward, engineers enjoyed a clearly identified social status accorded either by the state or by a collective recognition of the value and utility of their function.

Two models of engineering existed from the eighteenth century: that of state engineers and that of civil engineers who sold their advice to private entrepreneurs or occasionally to the state.

The interest of intellectuals in the practical needs of society was made evident in the sixteenth century and the opening of the seventeenth. It weakened at the end of the seventeenth century to the benefit of theory, because science had difficulty in supplying satisfactory answers to a multitude of questions posed by technology. This decline of interest was largely compensated by the development of state engineering schools, which provided a framework for the relationship of scientific learning and technological knowledge. In France, Saxony, and Prussia, state corps of engineers were created in the course of the seventeenth century. They were charged either with executing projects assigned by the state for fortifications, bridges, and roads or with controlling the output of mines and state manufacturers. In France, the first statute granted to a corps of engineers, created under Henri IV, was in fortifications. It was combined in 1691 with naval engineers to form a department of land and sea fortifications whose competence extended well beyond ballistics and the technology of fortifications, since it was involved with all realms of territorial management, cartography, urbanization, civil and military architecture, and hydrology. Vauban's works were a testimony to this versatility. Organized in 1716, the engineering corps of Ponts et Chaussées was charged with planning and managing the French networks of communications, roads, and canals. The authority of its engineers in the administration and construction of public works might overlap with that of military engineers.

In the eighteenth century, state-financed schools of engineering were created. The school at Mézières, founded in 1748 to educate military engineers, was "the most scientific of the technological schools of the eighteenth century."[1] Gaspard Monge taught there. The École des Ponts et Chaussées in Paris was founded in 1747. Besides stressing the importance of cartography in their instruction, these engineers were assigned to develop a planned system of water and road traffic by synthesizing existing technology. After the Seven Years' War, Saxony created an autonomous corps of experts in mining technology. First Prussia and then France, in 1782, followed this example. The science of mining had an ancient tradition that descended in part from salt mines controlled by the state. Based in mineralogy, it was the most prominent science of eighteenth-century Germany. Later, whereas the German mining engineers limited their sphere of expertise to mineralogy, French mining engineers were in addition assigned to supervise the development of steam engines, thereby continually extending their power and skills. Great Britain did not entirely reject the system of state engineers. During the sixteenth century, the English monarchy employed military engineers who played a major role in the technological evolution of land and especially sea armaments.

Thomas Savery, who in the 1690s invented a steam engine to pump water out of mines, was a former military engineer.

In the early modern period, another type of engineer emerged, modeled on Renaissance experts. These were generalists or specialists in one or another technological discipline, such as mines, armaments, metallurgy, glass, or construction. Some of them remained close to artisanship, whereas others became noted regional or international experts. Previously cited was also the example of British millwrights, who became true engineers in the second half of the eighteenth century. With special skills acquired in practice, they served both private industry and the state. This evolution was not confined to mills but concerned all of technology. Thus, the model of civil engineering stretched to sectors other than mechanics, particularly to public works and industrial equipment. The title of "civil engineer" was associated in 1771 with John Smeaton, the most celebrated member of the Society of Civil Engineers, who began his career by producing mathematical instruments with Henry Hindley but then greatly extended his skills. His objective in creating the Society was to differentiate civil from military engineers recruited by the state. A perfect example of a multi-talented engineer, Smeaton distinguished himself by constructing the new lighthouses at Eddystone, several bridges and canals, the viaduct of Newark, and the port of Ramsgate. He continued research on the mechanism of mills and on the Newcomen steam engine, improving the selection of materials and the size of pistons. While such engineering versatility was certainly not common, it never entirely disappeared, despite a progressive specialization that resulted from employment in enterprises and the application of university education to specific branches of knowledge.

The model of French mechanical engineers, whose title was recognized in the eighteenth century, was somewhat different from that of English civil engineers. Connected with artisanal practice, the former was derived from knowledge of materials and an analytic approach to technology. Artisans worked in contact with intellectuals and engineers who furthered the development of applied mechanics, for example, Étienne Lenoir, mathematician and producer of precision instruments; Edme Régnier, mechanical engineer; and Alexandre-Théophile Vandermonde, mathematician and director of the first office of public inventions. French engineers of the nineteenth century were the direct inheritors of this tradition in mechanics, which was institutionalized on the European Continent inside or outside of universities by specialized training in engineering sciences. Created in 1829, the central school of French engineering—properly entitled École Centrale des Arts et Manufactures—is the best illustration.

By achieving a synthesis between the empirical knowledge of workers and the formalized learning that they had acquired themselves, engineers were extending currents of action that were already apparent in the Enlightenment.

According to Antoine Picon, one may distinguish two stages in the technological thought of engineers. First, in the classical era, their activity was conditioned by "the doctrine of rational imitation" and by a vision of natural order that integrated technology. Then, beginning in the second half of the eighteenth century, they entered the path "of deliberate search for innovation," tending toward the promotion of change in order to assure the limitless progress of humanity and adopting a vision of nature and technology that was no longer ordered but instead founded on movement and fluidity.[2]

The Industrial Entrepreneur

At the time of the first industrial revolution, between 1760 and 1830, a new category of well-identified actors appeared: the entrepreneurs of industry, or simply "industrialists." These modern merchant-manufacturers were content to supervise in a general way the activities of production and to define their limits and specifications. They could provide employment to dozens or hundreds of domestic workers without being a true industrialist. Between them and the workers, an entire network of masters was interposed, ensuring an effective control of production. Industrial production was located on the periphery of their activities, whereas it was to be at the center for industrial entrepreneurs. Before the industrial revolution completely unfolded, one could find in Great Britain quite remarkable predecessors of true industrial enterprises. This was particularly the case in the metal industry, where societies in the seventeenth century were often created that concentrated numerous important sites of production directed by entrepreneurs capable of conceiving and introducing new technology. During the seventeenth century, enterprises and establishments of considerable size also developed in the paper and beer industries. In the eighteenth century, there were several examples of large establishments in sectors that remained otherwise dispersed, such as the factories of Matthew Boulton and John Taylor at Birmingham or the enterprises of patterned cloth. These were, however, isolated cases involving often fragile businesses.

The factory system did not really accelerate until the 1770s with the cotton industry, which produced the first important group of industrialists in the modern sense of the term. Richard Arkwright, the first true industrial entrepreneur of that sector, was its father figure. By 1835 there were 1,245 cotton factories in Britain. If these factories assembled only a few textile machines in the 1780s, by the 1830s they had already constructed large buildings to provide continuous production with the aid of specialized machinery. This development spread to other sectors in a dissimilar and irregular manner. In the iron industry, the use of coal allowed an increase in the size of installations and the formation of complex groups at different steps of the process, from the blast furnace to the rolling mill. In the chemical industry, it was not before the second half of the eighteenth century that real factories appeared. Hence, the industrial revolution did not reach its peak until the birth of a productive mechanical industry that used steam engines, machine tools, locomotives, and industrial machines. This sector became the heart of an interdependent network unified into a technological system.

The principal activity of owners of industrial enterprises was to supervise the manufacture of products in their establishments, which were now functioning factories. Devoting most of their time to management, they became "captains of industry" in the full sense of the term, as owners and operators of their enterprises. Like orchestra conductors, they were simultaneously capitalists, businessmen, and managers, but also very often technicians. In the textile industry they had frequently been salesmen and manufacturers, whereas in the metal and mechanical industries they were former engineers, workers, artisans, or technicians. Their role consisted of conceiving products based on demand; choosing the best materials, tools, and procedures; and lending the entire enterprise cohesion and impulsion by recruiting and training the best workers.

Some of these entrepreneurs utilized new techniques as a matter of routine, whereas others played a major role in their transformation. In an increasingly competitive world, the long-term survival of an enterprise depended on its capacity to improve products and procedures, either by introducing entirely new technological combinations or by ensuring their constant amelioration and adaptation to demand. In England and France, many entrepreneurs—in mechanics and metallurgy, but also in industries using machines, such as textiles—were themselves capable of devising such innovations because of their background as technicians. In all sectors they sought to reproduce a type of collective artisanal innovation, making the individual entrepreneur a major actor in the construction of knowledge. It is therefore altogether comprehensible that Joseph Schumpeter was justified, in his first model of development, to consider the innovative entrepreneur as the principal motor of industrial growth.

Science and Utility: The Other Revolution

Already in antiquity, the science of engineering, although clearly distinguishable from science proper, was oriented in its methods and principles toward knowledge of nature. In medieval times, university learning, which was derived from ancient knowledge, conserved for a long time an essentially philosophical character and included only a limited number of references to technological phenomena, even though they were a component of observable nature. Beginning in the thirteenth century, and especially in the sixteenth century, the dialogue between pure science and technological reality became modified by successive contacts that took different tracks from one discipline to another. The fruitful interaction of these two disciplines, technological and scientific, was taken for granted by the best intellects after the seventeenth century and became a general rule in the nineteenth century.

The dialogue between scientific and artisanal learning has thus been underway since the Middle Ages. The history of architecture offers a splendid illustration. The master builders and architects of the Gothic era used geometry to conceive their architectural forms. A major breakthrough occurred at the beginning of the fifteenth century: after rediscovering Euclidian geometry, Florentine painters and architects conducted research that laid the foundation for a science of perspective. To imagine the dome of the cathedral of Florence, Filippo Brunelleschi applied mathematical rules. The crown of the dome was constructed without scaffolding; it was formed by a complex fitting of superimposed pieces resting on interior ironwork. A work of technological prowess with a scientific basis, it united a perfect understanding of materials with a total mastery of forms, ushering in a complete revision of architectural practices. "The brilliance of Brunelleschi," writes André Chastel, "derived from his ability to reduce the structure to a mathematical rule and to untangle the play of elements that constituted it, which was informed by his notion of perspective."[3]

In *De re metallica*, Bauer moved technological knowledge a step further, insofar as it was doubtless the first encyclopedic undertaking to associate the technological object with an organizing principle. Several other treatises in the course of the Renaissance furnished in turn a concrete vision of technological processes. Discourses on agronomy, such as those of Charles Estienne (*La Maison rustique*) or Olivier de Serres, no longer settled for reproducing the work of ancient authors; instead, they advanced practical knowledge of farming and gardening. The work of Bernard Palissy proposed an equally concrete approach to chemical phenomena. This general revision of intellectual priorities during the Renaissance produced a serious reconsideration of artisanal practices. The simplification of knowledge in numerous realms—botany, zoology, anatomy, chemistry, geography—progressed on the basis of an attentive observation of nature that assumed an experimental character. Moreover, changes occurring in certain authors' vision of the world in the fifteenth century, as with the German Nicholas of Cusa, were confirmed in the sixteenth century by great discoveries in astronomy and geography, creating a new image of humanity's position in the universe. Science and technology were no longer simply a means to understand and imitate nature but actual instruments of its potential transformation.

Research devoted to technological change does not explain the origins of seventeenth-century science, but it was not absent in the scientific revolution. Numerous intellectuals, like Robert Boyle in 1663, asserted the utility of natural philosophy. Even if such utility did not principally concern technological applications, those were not excluded. This attitude extended to proposals by the

English philosopher and statesman Francis Bacon, who wanted to construct a science capable of affording humans a means to dominate natural phenomena and who even foresaw the creation of a collective organization uniting intellectuals, authors, and artisans to carry out that project.

The Royal Society, founded in London in 1660,[4] and the Academy of Sciences, founded in Paris in 1666, were charged with the explicit mission of developing knowledge useful for society and industry. During the early years of their existence, both organizations devoted much of their efforts to accomplish this task by examining such sectors as hydrology, pneumatics, and timepieces. Research into the means of determining longitude was the best example of these endeavors. The Royal Society conducted a project to record the history of commerce that fostered precise descriptions of industrial practices—for instance, the production of candles, parchments, dyes, beer, and salt—and thus participated in the process of formalization. It also developed a particular interest in mining and naval technology. One of its founding members, William Petty, wrote an essay in 1674 on the use of ratios in navigation, artillery, and construction. After the end of the seventeenth century, when it became clear that science could not improve practical results as easily as supposed, theoretical concerns superseded this scientific orientation, and curiosity replaced utility.

In France, this change followed the revocation of the Edict of Nantes in 1685, which caused the departure from the Academy of Sciences of foreign Protestants such as Christiaan Huygens and Olaus Römer, who had participated in its initial activities. The settlement of 1699 abandoned Bacon's ideal of collective efforts to gain utilitarian goals, and the balance sheet of such achievements drafted in 1739 was in the end rather modest.

The savants of the seventeenth century had hoped to construct a "natural philosophy" capable of fathoming the secrets of nature. Analyzing and understanding at least some of the functions of objects and of technological procedures were features of this program. But existing knowledge did not provide the means for such an understanding. The new sciences consequently developed independently of technology, both in content and in institutional structure. Nonetheless, the connection with technological phenomena was not broken, since the very mode of scientific inquiry was based on observation, experimentation, and a vision of natural order inherited from the Renaissance, all of which led to a deductive method. The scientific revolution of the seventeenth century thus depended in great part on the invention and deployment of scientific instruments adapted to experimental needs. This objective to improve knowledge through experimentation was the precise purpose for the founding of the Royal Society. Hence, the production and utilization of scientific instruments was not simply one of the most remarkable manifestations of artisanal creativity; it also became a crucial element in the construction of fresh knowledge based on experimental demonstrations. To achieve that goal, philosophers participated directly in the conception and production of these instruments, and their competence in this domain became integral to physics. Galileo thus took a direct part in creating the thermometer in 1595, the telescope in 1609, and the microscope in 1610. Huygens was initiated at an early age into the production of mechanical devices—along with geometry—and he constructed them throughout his life. Such was the point of departure that gradually imposed norms of precision on scientific instruments that were similar to those for industrial machines.

Since the beginning of the seventeenth century, the science of air pumps was one of the privileged realms of experimentation, being of interest to both philosophers and engineers. As Galileo had anticipated, it involved at the same time a study of vacuums, of atmospheric pressure, and of the elastic power of vaporized water. This research provided the theoretical and practical bases of an ensemble of knowledge that facilitated the conception of steam engines.

Several systems of knowledge in chemistry coexisted in the early modern era. Some of them were constructed in the trades using chemical procedures, such as metallurgy, dyes, and glassworks, which were handed down by apprenticeship from generation to generation. Since the sixteenth century, in some disciplines an initial stage in the formalization of such knowledge appeared with the publication of treatises describing procedures used and experiments accomplished. Simultaneously, in a highly polemical atmosphere, a chemical science was elaborated that was closely associated with metaphysics yet could not avoid interaction with artisanal and industrial practices. After the beginning of the seventeenth century, this theoretical chemistry became liberated from scholastic dogma. Generally speaking, one may say that the attainments of classical science radically modified perceptions of nature through an association with technological procedures. The mechanics of Galileo, the optics of Isaac Newton, the works on heat by Boyle, and many other advances jostled the existing vision of nature and thereby opened technological perspectives that were to be pursued by chemists of the eighteenth and nineteenth centuries.

The desire to place science in the service of industry and the progress of humanity gained redoubled force during the Enlightenment. It was in France that it found its most extensive field of application. The Academy of Sciences gradually renewed its utilitarian preoccupations. Since the academicians were leading experts in technology, their advice was solicited for technological projects and patent submissions. Maintaining their interest in mechanics and the maintenance of machines, they increasingly turned to questions of public utility and urban planning. In the years before the French Revolution, they essentially included industrial problems, metallurgy in particular, although without displaying much competence in this realm. Truth to tell, their interest in research of a practical sort had a rather ambiguous character. In 1783, Condorcet praised Henri Louis Duhamel du Monceau for orienting science toward utility, but he contended that the Academy should be principally occupied with theoretical science. Faced with increasingly numerous inventors, he said, the Academy needed above all to defend its province by insisting on its learned sponsorship of inventions.

This impetus was passed on in France and Germany by state engineers and in Great Britain by civil engineers. French savants, in the shadow of Gaspard Monge and Gaspard François Riche de Prony as graduates of engineering schools, of course entered the Academy of Sciences, but their research and discoveries were closely associated with professional activities. In France, intellectuals were educated and disciplines were constituted, with both boasting a dual allegiance to pure science and to "useful" knowledge. In the schools of engineering, at the École Polytechnique, and at the Academy of Sciences, a science of applied mechanics, which would radically transform the conception of machines, was thus constructed. In addition to being an increasingly efficient instrument of production, the machine was to become a conspicuous symbol of life and of reason. At the engineering school of Mézières, Monge laid the basis for descriptive geometry with a study of problems posed by trimming stone or wood and the construction of buildings. In 1767 it was another member of the engineering corps, Jean-Charles de Borda, who defined the maximum effect of flow on waterwheels. His conclusions had a direct effect on Lazare Carnot when he wrote *Essai sur les machines* and on Jean Victor Poncelet when, in 1823, he invented paddle wheels to equip the arsenal at Metz.

Practice and Theory

In the eighteenth century, the initially parallel paths of science and the chemical industry finally merged. Chemical experts multiplied their experiments, giving them an increasingly rigorous and quantifiable content, thanks to the utilization of scales and other high-precision measuring instruments. These experiments enabled them to discover several new gases and metals, such as nickel and platinum, as well as to conceive a genuine pneumatic chemistry, which, following the work of Joseph Black and Joseph Priestley, led to the theory of combustion elaborated by Antoine Laurent de Lavoisier in 1775. By defining the notion of elements through his experiments with air and water, Lavoisier established a new discipline that erupted in all directions during the nineteenth century and thereby created the necessary instruments of the chemical industry. Yet industrial chemistry was rarely the result of the pure and simple application of a new principle or of a new procedure imagined by theoreticians. Several authors have shown in recent studies that the dialogue between the empirical knowledge of artisans and the theoretical learning or experimentation of intellectuals was difficult and sometimes even contentious. In his investigation of the Belgian chemical industry at the end of the eighteenth century, Robert Halleux describes "the complex play of interactions between a developing academic chemistry and that of arts and trades." In general, Halleux asserts that no invention properly speaking was due to the progress of academic chemistry; rather, he concludes that innovations were "linked in the chain of improvements in traditional know-how, which was skeptical about academic learning."[5] However, at the end of the eighteenth century, several spectacular successes, such as new methods of producing sulfuric acid or chlorine bleaching, were the result of initiatives by savants like Claude Louis Berthollet who were persuaded that the future belonged to the direct application of theoretical science to industry.

Chapter 4

Technological Adventures

A brief reconstruction of three "adventures"—in the sectors of energy, mechanization, and chemicals—should allow a better estimation of changes in acquired knowledge during the eighteenth and nineteenth centuries. Mastering the transformation of heat into motion by steam engines, producing machines that transmitted power and transformed motion without human intervention, and finally creating a chemical industry constituted three of the major technological breakthroughs that defined the first industrial revolution.[1]

The Production of Energy

Until the middle of the nineteenth century, the proportion of hydrological energy in Europe, and in the United States as well, constantly increased to become in the 1850s the principal source of motor power, more than that of humans or animals. This history had two successive phases: the adoption of diverse solutions to utilize energy and the adaptation of that power to devise new procedures.

Hydrological Energy

Since the end of the seventeenth century, a science of mills, designed to obtain better results, had been developing through modern empirical hydrology. This opposed a theoretical approach of French origin to a strictly experimental approach that was Anglo-Saxon. In the seventeenth century, mathematicians like Edme Mariotte analyzed problems posed by mill wheels without reaching a consensus. The experimental approach was introduced in the following century by the English engineer John Smeaton and his disciples, who worked on the submerged wheel, and by the brothers Austin and Zebulon Parker, who invented wheels simultaneously using impulsion and reaction. The size of wheels was considerably increased, and iron began to replace wood. Several intellectuals in France followed Mariotte's lead, and in the first third of the nineteenth century the French succeeded in combining the two approaches: Jean Victor Poncelet demonstrated the superiority of curved paddles on mill wheels, and Benoît Fourneyron invented the turbine. The turbine was one of the first examples of a machine conceived entirely through theoretical reasoning. Adopting Nicolas Sadi Carnot's principle, Fourneyron calculated the dimensions and defined the optimal forms of curved mill wheels, the size of the wheels, and the volume of water. He molded paddles to make them convey water without a shock. Perfecting this invention required many years.

In the United States, the manufacture of mills was still a dominant feature of the mechanical industry at the end of the eighteenth century. In 1795 Oliver Evans published a book entitled *The Young Mill-Wright and Miller's Guide* in which he combined a concrete descriptive approach with a simplified presentation of current French theories. He suggested that actual experience with the movement of wheels and water was more significant than theoretical deductions, and hence theory was an imperfect guide for constructing mills. Such prejudice dominated American hydrology until the middle of the nineteenth century, and it was not before the 1860s that engineers replaced millwrights in large factories. Between 1849 and 1855, following the principles of Fourneyron, James Bicheno Francis invented an "American turbine," which was adapted to moderate flows of water as a result of experiments combining scientific knowledge with practical methods. The science of liquids was not yet advanced far enough to offer a response to questions posed by phenomena such as friction. It was not until the end of the nineteenth century that the means were developed to do so, fostering the rise of centralized hydro-electrical plants.

Steam Energy

The Newcomen steam engine underwent several improvements after its invention by Thomas Newcomen in 1712. Especially notable were the injection of cold water at the instant when the piston reached the peak of its movement, thus increasing speed, and the automatic control of water valves, using a camshaft to agitate them, which produced greater regularity. Smeaton conducted research to optimize the choice of materials, the diameter of cylinders, the size of pistons, and the means of injecting cold water. As a manufacturer of instruments, he worked in close collaboration with the University of Glasgow, where an eminent specialist in the properties of heat, Joseph Black, was a professor. James Watt's first model of a steam engine was conceived in 1765 at the request of John Roebuck, owner of the Carron foundry in Scotland, who was encountering difficulties in pumping water from his coal mines. Watt had already recognized the weaknesses of the Newcomen engine and sought to correct them. Following Black's idea about the latent heat of vaporization, he found a solution by eliminating the direct injection of cold water, which caused a huge waste of vapor needed to reheat the cylinder that had cooled during the preceding phase of condensation. The vacuum thereby created beneath the piston was much greater than with Newcomen's engine. Before delivering his first machine in 1776, Watt had to overcome a considerable number of obstacles. After Roebuck went bankrupt in 1773, Watt partnered with Matthew Boulton, who had created at Soho (Birmingham) a large enterprise that manufactured ornamentation and household items.

During the thirty years that followed, Watt and other engineers attempted to design a machine capable of producing a rotary motion so that it could be used in industry and not solely to lift water. In 1779 and 1780 Matthew Washborough and James Pickard vastly improved the mechanism for converting straight motion into circular motion.[2] Unable to obtain their patent, Watt designed another machine of great complexity. The main improvement realized by Watt in 1782 was a machine that, besides employing a condenser, alternately emitted steam above and below the piston, allowing the steam to agitate four valves successively. Thus, the machine became automatic. Watt's invention was a collective effort par excellence, created through a synthesis of many new ideas constituted from artisanal know-how and scientific knowledge. All those ideas could not have emerged from a single brain.

The Mechanization of Industry

Artisans as well as domestic workers very early used mechanical procedures that were at first no more than accessories to tools. Under the pressure of growing demand, such mechanization accelerated in the course of the eighteenth century. It developed initially within the traditional framework of technology based on the use of low-power energy. The mechanical objects employed by artisans and workers in workshops became gradually transformed into actual machines. During the years from 1760 to 1780, mechanical innovations appeared in the use of hydrological and steam energy, as well as in the means of transmitting motor force. They spread into numerous sectors of activity, the most symbolic among them being the textile industry, which became the leading sector of the industrial revolution.

Mechanization of Textiles

Even before the eighteenth century, many mechanical systems were introduced into the process of textile production. Among them were the pedaled spool, displayed in a celebrated painting by Velázquez, which allowed the spinner to better guide and twist thread; the craft of Dutch weaving; and the knitting machine of William Lee, which appeared in the sixteenth century. The patent taken for a flying shuttle by a Lancashire artisan who manufactured combs, John Kay, dated from 1733, and that of Richard Roberts for a device of automatic weaving from 1825. In the course of nearly a century separating these two dates, various radical innovations that profoundly modified the methods of work and significantly increased productivity ensued in the cotton and drape industries. This wave of inventions has for a long time been explained as a game of challenge and response between spinning and weaving, with an advance in one sector revealing an unbearable gap of productivity in the other, thereby creating a bottleneck, which, due to a strong increase in demand, provoked research and the adoption of a new procedure in the lagging sector. This interpretation has been justifiably contested because it does not stand either a precise examination of the facts or a statistical analysis of innovations tabulated by historians and confirmed by applications for patents.[3] Kay's flying shuttle did not produce a significant increase of innovations and applications in the years following his invention between 1734 and 1753. In reality, the use of the shuttle did not truly develop in the cotton industry until after 1760. The principal reason for its adoption was not the search for increased productivity—for small pieces of cloth, the increase was quite feeble—but the desire to improve the quality of cloth in order to respond better to customer demand. Even in the years 1760 to 1799, no increase in the number of innovations and patent applications in cotton spinning could be observed. In fact, the most creative sector of that time was the hosiery trade.

In 1767 James Hargreaves invented the spinning jenny, which was formed of spindles affixed to a chassis and was operated with a crank turned by hand. Imitating the gestures of a spinner, it was used in domestic industry. In 1769 Richard Arkwright constructed his water frame, which could be activated by a water mill and which stretched thread with a draw roller. It improved the spinning operation by eliminating manual operations and utilizing mechanical parts. In the 1790s machines appeared with a hundred spindles. Nevertheless, during the years 1765 to 1784, the distribution of innovations and patent applications for spinning, weaving, and finishing cloth was not radically modified, as one would have expected if the hypothesis of challenge and response were valid. By contrast, innovation continued in spinning. The threads produced by the spinning jenny were of inferior quality, while those furnished by the water frame were rigid. These difficulties were overcome in 1779 by a machine of Samuel Crompton, which combined the stretching mechanism of Arkwright with the alternating device of Hargreaves to permit unlimited variations of speed. The chief motive of innovation was again the pursuit of quality. Improvements in spinning continued to mount after 1775 when Arkwright applied for a patent to join carding and spinning into a single process.

In weaving, the first patent for a mechanical device was applied for in 1779 by William Cheape, followed in 1785 by the patent of Edmund Cartwright. In the next decade and a half, sixteen improvements were added, half of them patented, without the slightest acceleration of innovation in spinning. Cartwright's machine did not catch on until after 1800, and then not without undergoing improvements. In 1813 the trades involving hand production were still ten times more numerous in Great Britain than the mechanical trades. With the invention of an automatic machine by Richard Roberts in 1825, the initiative in innovation passed into the hands of mechanical engineers. The great specialized mechanical industry was born. All in all, the mechanization of the textile industry does not seem to have been the result of disequilibrium between the levels of productivity in spinning and in weaving but was oriented instead toward an effort to adapt products to the market.

Mechanization in Other Sectors

In addition to textiles, other sectors such as pins and paper underwent a process of mechanization at the turn of the nineteenth century. The mechanization of pin manufacturing began with a division of operations into a dozen successive steps. After some unsuccessful attempts, a manufacturer of mathematical instruments patented a machine in 1809. Although it was only partially mechanized, its use spread. Complete mechanization was not accomplished until the 1830s when a method was created to manufacture a pin with both head and stem in one piece.

The mechanization of the paper industry extended the previous trend of this sector to develop continuous modes of production. In the eighteenth century, considerable progress had been realized in the preparation of paper pulp as a result of the invention and improvement of cylinders. But thereafter, between the preparation of pulp and the formation of paper sheets, an increasingly serious "distortion" in the manufacturing process became evident.[4] A bottleneck was created that had to be surmounted in order to increase production and respond to demand. The first patent for a paper machine was obtained in 1799 by Nicolas Louis Robert, a mechanic who worked at a paper-making factory in Essonnes. For this achievement he was awarded a prize of 3,000 francs by the École des Arts et Métiers. This procedure, along with improvements, spread rapidly after 1800 in both France and Great Britain. Robert's machine represented a true technological breakthrough, since it did not depend on a reproduction of gestures by workers but executed its functions without imitation. Instead of raising sheets of paper one by one, thus risking tears, it produced continuous rolls of paper after passing them without pause along a wire netting. These machines, called "flat tables," quickly outdistanced other models. Applied in other industries like chemicals and distilleries, these methods were expanded after 1850 to include the production of foodstuffs and quotidian objects.

The Birth of the Mechanical Industry

Between 1760 and 1840 a specialized mechanical industry was created that, in the 1840s, furnished all types of industrial machines as well as steam engines and mechanical tools. This industry was the fruit of a long process of fusion among various sectors of both empirical and tacit knowledge realized by artisans and workers—mill builders, carpenters, blacksmiths, metalworkers, tool manufacturers—and by engineers or technicians who themselves grew out of a working milieu. The integration of artisanal knowledge into a global mechanical synthesis rested on a dialogue between emerging formalized mechanics and manual skills. Very often the machine simply reproduced a worker's gesture, but this gesture was submitted to an increasingly rigorous demand for precision. Mechanical precision, which was developed quite early in clock-making and scientific instrumentation, opened the way to a general mechanization. The production of steam engines, boilers, and textile machines required the use of ever more accurately measured components. The response to this demand was provided by a more elaborate conception of mechanical tools, which thus became the principal vector of general mechanization.

The dispersed world of mechanics became unified, integrating into a single ensemble the different specialties and forms of knowledge organized around the industry of mechanical tools. This transformation followed different paths from one country to another to arrive at an identical result. The mechanical industry of Paris developed from precision mechanics, originally in clock-making and scientific instruments, to the creation of industrial machines, first in textiles, and then to mechanical tools and steam engines.

Between 1770 and 1830 a strong mechanical industry also developed in England. It passed from the production of hydrological machines and mills to steam engines and other mechanical devices—everything from textiles and sweets to paper products—before arriving at the manufacture of all sorts of mechanical tools and locomotives. This was the work of engineers and entrepreneurs who knew how to compose fresh knowledge from their artisanal experience and simultaneously to reconstitute an entire system of skills to meet the needs of these new technologies, particularly the demand for precision. The lathe tools of John Wilkinson, invented in 1775, made possible the use of steam-driven cylinders. Henry Maudslay, a manufacturer of tools, constructed spindles that were instruments of great precision. Progressively, industry moved from a world where work was executed by hand to a world where these operations were performed by machinery. Without the slightest discontinuity, artisanal skills thereby created tools that offered the possibility of transposing manual know-how into a world of mechanics.[5]

The history of Boulton and Watt's Soho Foundry well illustrates this transition, just as it demonstrates the central role played by industrial entrepreneurs in the accumulation of knowledge. Built in 1795 to produce steam engines, it was the first large factory of mechanics. Not content to attract the qualified workers he needed, Watt selected specialized groups among them that he taught to meet his needs, for precision in particular, encouraging them to have their sons join as apprentices. Hence, as a biographer of the famous millwright William Fairbairn indicated in 1877, "the facilities offered brought a regular progressive improvement in the nature of work accomplished and meanwhile a constant reduction of dependency on purely manual labor."[6] One cannot therefore separate the rise of mechanization from the evolution of know-how. Many of the Soho workers became highly qualified technicians and moved out into other factories. The expertise necessary for this new technological system to function was controlled by master mechanics coming from mill construction or by artisans and workers who became real engineers and inventors, of which there were four or five hundred in London by 1825.

The rise of the British mechanical industry thus depended largely on the development of empirical knowledge. The same was not true in France, where, on the basis of theoretical mechanics, applied mechanics appeared and constituted one of the principal subjects taught at the École Polytechnique and, some years later, at the École Centrale.[7] In the United States, the publication in the late eighteenth century of Evans's book, *The Young Mill-Wright and Miller's Guide*, marked the first step in a process of formalizing mechanical knowledge, which was to make of the US a half-century later, at the time of the Great Exhibition of 1851 in London, the leader of this industry.

The Chemical Industry

The dialogue between chemical science and industry began in the eighteenth century. It resulted, among others, in three remarkable accomplishments: a new method of producing sulfuric acid, the bleaching of cloth by chlorine, and the Leblanc method of producing soda ash (sodium bicarbonate). The production of sulfuric acid, used in the dyestuffs industry to dissolve indigo, was still negligible at the opening of the century, but its use developed in spite of procedures that allowed only small quantities to be derived. In 1746 Roebuck invented a lead chamber, based on a new welding procedure, and started a business firm in 1749. Manufacturing in the chemical industry passed without transition from a stage of artisanship to that of large industry, and between 1770 and 1780 factories spread throughout Europe.

Bleaching cloth sheets was a complex operation that required time—four weeks—and was undertaken only in winter. It demanded a large quantity of water, vast space, and the use of soaps and various acid baths. After 1750 several chemists attempted to improve this procedure. In 1773 the Swede Carl Wilhelm Scheele showed the destructive power of chlorine for dyes, and in 1785 Claude Louis Berthollet demonstrated the possibility of dissolving gas and using this solution for bleaching. The adoption of this procedure was slow, however, as it was long accused of weakening the solidity of fabrics.

The primary materials used to produce the ammonias necessary for the industries of soap, glassware, saltpeter (potassium nitrate), and bleaching were composed of vegetable powders. For soda ash, a plant called barilla, principally from Spain, was employed. In 1737 Duhamel du Monceau uncovered the possibility of producing soda from salt. Six processes in all were introduced in England and France, but because they were too complex, their productivity was limited. The right solution was found by Nicolas Leblanc—the private physician of the Duke of Orléans as well as a chemist and surgeon—who used the reaction between salt and sulfuric acid. This process, despite approval from several intellectuals, was not supported by successive French governments, but it was very well-received in England, where it rapidly developed.

These three products allowed a response to expanding industries that were as varied as printed cloth, soaps, and bleaches. Accordingly, society became more preoccupied with hygiene and more inclined toward colorful garments. Each of these advances was the consequence of a dialogue, sometimes boisterous, between academic intellectuals and tradesmen possessing empirical knowledge. All were rooted in traditional know-how that induced reflection and reconciliation with theory. These developments owed much to new chemical theories and laid the foundation of industry by exploiting the discoveries of mineral chemistry.

Thus, the question is posed of the relationship between those possessing empirical and tacit knowledge, derived from artisans and workers, and those representing academic knowledge in a broad sense of the term. It was the fusion between the two, progressively realized in the nineteenth and twentieth centuries, that became the principal source of technological change in the modern era. This process developed thanks to an approach of collective innovation founded on the dialogue and exchange of experiences between the advocates of these forms of knowledge. One should therefore no longer look on empirical knowledge, transmitted from generation to generation, as being opposed to the innovative practices that resulted from the application to industry of formalized and codified knowledge. It was the fusion between them that transformed the Western world through the industrialization of the nineteenth and twentieth centuries.

Part Two

The period extending from the 1830s to the 1960s saw the formation of a system of constructing scientific knowledge and technological practice that was inspired by the values inherited from the European Enlightenment. It was at the same time a question of understanding the universe better and of realizing the well-being of humanity. The institutional framework of this development included an ensemble of actors who possessed both scientific knowledge and a closely related technological culture. These were the sources of innovation.

The great protagonists of this story were artisanal trades, engineering, entrepreneurship, and universities. Artisanship had enjoyed a rapid evolution in early modern times. It contributed to the formation of engineers and was itself transformed in the process. Engineering capability was derived from experience in the daily conduct of technology on the basis of education received at engineering schools and universities. Enterprises were natural meeting places for those who elaborated technological learning and models of organization. They encouraged specific knowledge by integrating activities of research simultaneously and by cooperating with external actors, principally from universities. University disciplines were organized in rigidly hierarchical networks that applied rigorous procedures, thus assuring the coherence of the academic community. In its various incarnations, the state intervened in the construction of knowledge through the major role played by science and technology in championing nationalism.

After a general presentation in chapter 5 of these different aspects of technological change, we shall examine two illustrations of these mutations: the history of the steam engine in chapter 6 and that of organic chemistry in chapter 7.

Chapter 5

Institutional Logic and the Dynamics of Knowledge

Since innovation is essentially collective, a technological object is the product of a fusion of different types of knowledge, in both its conception and its use. An analysis of the formative process of technological knowledge in workshops, factories, and laboratories must therefore take into account the diversity in the concrete experience of the actors—entrepreneurs, artisans, workers, engineers, and scientists. All of them acquired a different education and belonged to different learned communities and social groups at the interior of which diversified cultures and social practices coexisted. The encounter among these various cultures created inevitable tensions. However, only the capacity of individuals to surmount these differences and to combine forms of knowledge could create innovation and fresh learning. Hence, the necessary convergence of knowledge could not be realized except in certain privileged locations that encouraged dialogue. Rather than contrasting the different types of knowledge, as is so often the case, historians of technology should attempt to understand why and how such locations of convergence appeared and became centers of creativity.

Between the 1830s and the 1960s, a mode of constructing knowledge formed that brought into contact four quite distinct spheres of activity: enterprises, trades, engineers, and academics. Each of these communities possessed a dynamic of creating knowledge that was separate but that shared relations with the three others, which doubtless gave a decisive impetus to creativity. This mode of construction, combining specialties of knowledge with their permanent interaction, gradually came about during this span of time, attaining full stature in the 1950s.

Starting in the 1880s, large enterprises emerged that developed planned and collective research as well as strategies for appropriating knowledge principally by acquiring patents. They thus became locations for accumulating knowledge that constantly needed to be maintained and renewed in order to survive. Increasingly better educated and organized, engineers elaborated specific modes of constructing knowledge within the framework of well-defined disciplines, forming a large corpus of engineering science. This was the result of a synthesis between the practice of trades, original research, and knowledge borrowed from pure science. Meanwhile, the academic community developed in a more rigid fashion around a curriculum founded on the separation of hierarchic and homogeneous disciplines. Academics professed an intention to pursue a program of pure science, but they gradually became involved in the development of useful or practical knowledge. Arrogant claims for the superiority of fundamental science over applied science slowly lost credibility because technology itself became a ubiquitous component of the real world.

In reality, at the turn of the nineteenth century, the dialogue between empirical and academic learning was the source of frequent tensions that were often conflictive, yet the process of combining different branches of knowledge during the century became the basis of collective innovation. One might think of the triad formed in the 1850s at Glasgow by William Rankine, John Elder, and William Thomson, or the team assembled by Thomas Edison in his laboratory at West Orange, New Jersey, in the 1880s. These instances were not incompatible with that of individual inventors who dominated the century. In some sectors, the combination of pure science and practical technology was achieved very early, but in others it was not attained until after a long and difficult process. It was not rare to observe two systems evolving in a parallel rather than complementary manner before becoming compatible and unified.

Entrepreneurs and Enterprises

As a basic premise, the enterprise may be considered as one of the privileged sites of innovation and the construction of new technological and scientific knowledge. This assumption recognizes the fact that solely an enterprise, whether private or public, is capable of transforming scientific learning and technological know-how into useful products and consumer services. The prominence accorded by historical literature to the relations between enterprises and universities is a testimony to that fact. Entrepreneurs cannot assure the survival of their business without keeping pace with the most advanced technology. They can do so only by innovating, by combining their knowledge with that of close collaborators, and by integrating knowledge obtained elsewhere—from universities, engineering schools, or other enterprises.

The Creation of Enterprises

It was by responding to growing consumer demands that enterprises were created in the nineteenth and twentieth centuries. Under the pressure of competition, some of them became a permanent source of innovation. Whether their activity was devoted to producing or advising, these enterprises were points of convergence for technological know-how. The success of innovative enterprises depended on their organization and their own experience, but also on the quality of skills that they were able to muster and the networks that they could assemble. It is therefore not surprising to see the creation of enterprises figuring in the history of economic growth and technological change and thereby defining the contours of our civilization through innovation.

The story of the first wave of entrepreneurial creations in France between 1800 and the 1870s is particularly enlightening in this regard. It was contemporary with the diffusion of technology in the first industrial revolution and was associated with the construction of large transportation networks on land and sea. In the case of the iron industry, the tentative and experimental character of such creations was clearly apparent. In the 1820s, several attempts to adapt English procedures ended in wretched failures, such as the Mining Company of Saint-Étienne, founded in 1818 by the mining engineer Louis de Gallois. In 1842 Eugène Charrière, who became director of the Compagnie d'Allevard after its collapse, succeeded in resuscitating the enterprise by orienting it toward the production of steel rails.

In the textile industry, the large enterprises created at mid-century had the objective of launching new products and democratizing their use through mechanization. Each French textile center thus acquired a distinct personality defined by the nature of its products and the procedures adopted. In the sector of luxury items, enterprises such as Baccarat (crystal), Christofle (silver), and Cartier (jewelry) owed their success to mastering innovative technology used for new products. In all these examples, the primary condition of success resided in the internal coherence of the initial innovation due to various new forms of knowledge and a close connection between the skills and know-how necessary for development.

Another wave of innovations, arriving with the second industrial revolution in the 1880s, was also the result of initiatives by new entrepreneurs rather than enterprises already in place. In electricity, the chemistry of synthetics, and automobiles, these new creations reflected the aspirations of a society in the midst of a transformation that was perceived more or less clearly by the founders of enterprises. These leaders depended on precise programs of action defined by an understanding of technical systems, the expectations of consumers, and the present or future opportunities offered by the state of technology and science. Enterprises were essentially created to conceive and manufacture new consumer products that incorporated basic technological and scientific knowledge of modern civilization—electronics, electrochemistry, organic chemistry—in order to produce, for example, the internal combustion engine, the telephone, and television. Their founders were usually independent inventors, such as Paul Héroult, Leo Baekeland, Alexander Graham Bell, and Hermann Hollerith, who conceived the projects and assured their development.

The best example of the dynamics of innovation tied to the initiative of new enterprises is that of the automobile and aeronautics industries in France. One must recall the initial modesty and uncertain prospects of the projects started by the Michelin brothers Édouard and André, by Armand Peugeot, or by Louis Renault. Unlike enterprises that began with a captive market and a practiced technology, most of the companies founded in this era undertook the adventure of imagining new products. Emmanuel Chadeau has shown that the first aviation workshops were established even before there was a market for airplanes.[1] Many of the new entrepreneurs were also inventors with an engineering education. In France, the École Centrale and the Écoles des Arts et Métiers were seedbeds of enterprise. For the most part, they were integrated into a network where knowledge easily circulated, giving to the creative process a collective as well as an individual character. Men such as Zénobe Gramme and Georges Claude in France and Edison and George Westinghouse in the United States were immersed in the techno-scientific milieu of their time.

Thus, clusters of startling innovation were formed around new technologies, for which Paris of the Belle Époque provides an ideal model. It was a place where inter-enterprise relationships were intense and where men, knowledge, information, and capital circulated without difficulty. The spectacular success of Paris and its suburbs, as well as that of Berlin at the turn of the twentieth century, was the result of strong interaction among enterprises that in no way excluded their lively competition. The innovative capacity of small and medium enterprises that had plunged into this atmosphere was also exemplified in numerous sectors of the American mechanical industry, such as hydrology and machine tools, whose increasingly precise, diverse, and automatic products entered the world market after the 1860s. They were constructed and perfected by medium-size enterprises that specialized narrowly in a limited number of products closely related to one another and that employed knowledge specific to each of them.

This model of entrepreneurial creativity likewise applied to several innovative sectors of the third industrial revolution since the 1960s, principally marked by the rise of information technologies. Giovanni Dosi has described how the development in the 1950s and 1960s of the microprocessor industry in the United States was the result of initiatives by some small firms "with entrepreneurial dynamic" rather than by larger enterprises in the electronics sector. "The history of commercial success of new products," he writes, "shows the primordial role played by new companies (usually emerging from existing companies producing semiconductors)." The author explains their enduring role by the fact that, in this sector, the technology of the times was "embodied in persons rather than equipment."[2]

This remark is valid for most of the innovations appearing in information technology since the 1960s. They were small or medium-size enterprises created by men who, after their university education, had often worked in research laboratories of large enterprises. The Bell Laboratory founded by AT&T, which held a monopoly on telephones in the United States, was the first among them. Their teams of researchers were composed both of scientists and men capable of mastering the available technology and following its evolution.

The Enterprise as a Convergence of Knowledge

Having overcome initial obstacles, the survival of an innovative enterprise depended on its capacity to construct new knowledge and to assimilate outside information rapidly. It thus became above all a site of accumulating knowledge in order to form its identity and to ensure its longevity in a competitive economy. Hence, the enterprise essentially depended on the capability of its directors and employees to engender fresh knowledge through a constant effort of innovation. To attain that goal, the enterprise had to coordinate different skills and aptitudes in order to meet the needs and expectations of the market. The creation and activity of an industrial enterprise led to adventures more collective than individual. Success was the result of teamwork that united complementary skills. These generalizations apply equally as well to the small or medium-size enterprises that dominated the economy in the nineteenth and early twentieth centuries as they do to the large ones that flourished in the later twentieth century.

One frequently cited example of a successful collective innovation is the Edison Electric Light Company, founded in 1878. It was the fruit of well-directed teamwork that united an intellectual, a skilled technician, several artisans, and a businessman versed in public relations. The same may be said of the successful Westinghouse firm in the 1880s. Likewise, a close study of innovative enterprises in France discloses among their founders and close collaborators the presence of various professional groups: businessmen and financiers, who knew the markets and banks; artisans and workers, who possessed the know-how required for manufacturing; and especially engineers and technicians, who emerged with increasing frequency after the 1840s from technical and engineering schools such as the Écoles des Arts et Métiers and the École Centrale.

Many of them acquired practical knowledge from artisans and workers, and they possessed formalized learning founded on engineering science and theoretical study. Increasingly, among founders of enterprise and their advisers, numerous university scientists appeared, briefly or permanently abetting the enterprises. They played a major role in the first half of the nineteenth century, particularly in Great Britain and France. Henri Victor Regnault and Henri Sainte-Claire Deville are splendid examples. The former became a professor at the Collège de France (1841) and director of the porcelain factory at Sèvres (1854); during the Second Empire, he was also chief analyst in the laboratory of the Compagnie du Gaz de Paris. The latter developed a process for the industrial production of aluminum early in the 1850s and participated in creating two chemical companies that used his design to help develop what became the Pechiney firm. German chemical enterprises founded in the 1860s, such as Bayer, Hoechst, and BASF, created a giant industry of dyes and synthetic medicines that dominated the market prior to the First World War by bringing together highly trained chemists, technicians skilled with vegetable dyes, and businessmen familiar with the market.

Meanwhile, along with individual entrepreneurs and medium-size firms, large enterprises participated in creating new products and new sectors of activity. That fact is well-illustrated in France by companies like Saint-Gobain in glassware and Schneider in metallurgy. These enterprises maintained close ties with intellectuals such as Nicolas Clément-Desormes, a professor at the Conservatoire National des Arts et Métiers. Moreover, very early on, certain enterprises created laboratories to test chemicals or machines designed to improve the quality and measure the performance of products or materials. Such laboratories, which were widespread in the chemical, iron, and steel industries, responded to the needs of an increasingly demanding and powerful clientele, whether large civil and military administrations or public service organizations like railway companies or water and gas firms. Naturally, their evolution resulted in some disruptions within the technological system. The poor quality of materials or the faulty design of machines provoked spectacular and often fatal accidents—the explosion of boilers, the separation of railway tracks, or the collapse of bridges.

Consequently, as the editor of a technical encyclopedia wrote in 1886, "the laboratory has become a necessary auxiliary of all industries, especially at a time when through competition the old manufacturing brands have lost their former luster that previously reassured customers.... It is difficult to think of an industry that employs a product without testing it."[3] The railway companies, for example, installed inspection workshops for their suppliers of materials and products like coal. Many enterprises were obliged to start their own research laboratories during the century. Thus, in the United States, according to statistics of the National Research Council, 112 laboratories were created before 1899, 182 between 1899 and 1908, and 371 between 1909 and 1918.

This tendency was favored by the development of large enterprises and the evolution of business rights after the 1860s. Enterprises sought to limit the risks associated with technological innovation through a system in which the paths of technological practice and scientific thought were slowly converging. As a consequence, a model developed after 1870 of knowledge based on cooperation between universities and enterprises, particularly in the sectors characteristic of the second industrial revolution. By the 1880s, the concentration of enterprises was accelerating, especially in certain new sectors like chemistry and electronics. Enterprises such as BASF, Bayer, and Hoechst in the German chemical industry or Westinghouse and General Electric in American electronics implemented permanent policies of innovation founded on the development of integrated research and organized cooperation with academic circles. In a word, they internalized research so as to renew constantly the supply of knowledge that they possessed, controlled, and deployed. This strategy of integrated research was linked to a conscious policy of appropriating knowledge. Assured by the results of research, access to new knowledge was thereby secured for a time sufficient for its exploitation.

The creation of industrial laboratories between 1880 and 1914 thus constituted a basic element of business that might be called "vertical integration." Some of these laboratories extended their activities of testing and measurement into industrial and even scientific research. The birth and development of French enterprises of metallurgy in the nineteenth century deserve mention, since, starting from an empirical base, metallurgy opened out into metallography and metallurgic physics. Enterprises in the center of France, threatened with failure because of competition from Lorraine after the invention of the Thomas-Gilchrist process, intensely pursued laboratory research after 1870. They could survive only by "making a scientific study of metal properties to draw conclusions capable of guiding production," thereby "specializing in products of superior quality to create new outlets."[4]

Engineers, especially from the École Polytechnique and the École Centrale, were recruited by these enterprises to organize and direct their laboratories. One of their main activities was to perfect new methods for analyzing the quality of metals. In this way, they not only improved the technology and production of alloys but also participated in the evolution of metallurgical theory. For example, after having worked with Schneider, Floris Osmond and Jean Werth published in 1885 an article about the properties of steel. Osmond then left Le Creusot and independently pursued his research, becoming a widely recognized leader in that discipline. In a Montluçon factory of the Compagnie des Forges de Châtillon-Commentry et Neuves-Maisons, Georges Charpy considerably advanced the use of measuring devices, which he integrated into manufacturing. At Commentry-Fourchambault, on the initiative of Henri Fayol, the laboratory was equipped in 1912 with sophisticated instruments "to investigate special metal alloys in order to discover and utilize their unique properties."[5] Chemical metallurgy was born in these laboratories, which gradually became sites of fundamental research where theories and concepts were elaborated that accounted for the physical transformation of solids.

Most of the industrial laboratories in the United States were also initially interested in analyzing materials and controlling quality. They began by orienting research toward improving essentially practical projects related directly to production. Then they became interested in subjects of a clearly scientific nature, although it was considered feasible "to discover new areas of profitable activity," as the executive counsel of General Electric put it in 1900 to justify the creation of a research laboratory. This research would allow a diversification of products "by protecting established markets and creating opportunities for new ones."[6] The development of the laboratory demonstrated the close cooperation between fundamental research and engineering science that was oriented toward precisely defined needs of production. Such an orientation was not accidental but rather the result of a strategic choice. Elihu Thomson himself, in a letter to the president of the company in 1900, asserted that if General Electric wished to continue "to explore and develop new territories," it would need to have "a research laboratory for commercial applications of new principles and even for the discovery of those principles."[7] GE thereby readily passed from commercial to fundamental research. In 1909 the firm recruited an eminent intellectual, Irving Langmuir, who won a Nobel Prize in 1932. Langmuir's initial research led to the invention of a tungsten lamp that became a major source of profit for the enterprise between the wars, while his work on vacuum tubes opened new prospects for the theory of matter and laid the basis for industrial electronics.

The creation of General Electric's laboratory was the result of a slow recognition by directors of the American electronics industry in the 1890s that only "a research devoted to fundamental questions posed by materials and procedures" would permit them to counter technological advances abroad, particularly in Germany, which threatened the commercial position of America.[8] The creation and development of the GE laboratory occurred in the context of a general movement among enterprises that peaked before 1914 and that included DuPont (1911), Eastman Kodak (1912), and AT&T (1907–1910). All set up laboratories in response to two necessities: to meet foreign competition on an equal footing and to secure patents in order to protect established positions.

During the same era, according to Jürgen Kocka, research in the giant German enterprises began "a long-term process financed by means of increased investments." Such research did not always have an immediate or specific goal, such as the development of electric lighting or a new dye, but it might be oriented toward the diversification of long-term beneficial activities of the firm. "The case of the individual inventor," writes Kocka, "who achieved results without a supporting institution and then either sold his idea to a firm or established his own concern to exploit it himself … became less important, though it still occurred frequently."[9] The most remarkable example of this process in Germany was the creation of a laboratory in Elberfeld in the 1880s by the director of Bayer, Friedrich Carl Duisberg, at a total cost of a million and a half marks. The personnel recruited were essentially university-trained, and the research conducted there laid the scientific and technological basis for synthetic drugs.

Even though these laboratories developed research belonging more to engineering than to pure science, none of them could ignore the state of scientific knowledge, including the most theoretical, in the disciplines concerned with the activities of enterprise. They maintained a steady watch on the zone of contact between these two types of knowledge. All had the same mission, as defined by Elihu Thomson in 1900 for the GE laboratory: to improve products and procedures or to create new ones by profiting from eventual discoveries belonging to engineering and theoretical science, whether conceived in the laboratory itself or elsewhere at universities, other enterprises, or consulting firms. In these large laboratories, research was programmed and planned in a manner often more rigorous than in university laboratories. Doubtless, the most characteristic example was the success in 1909 of Fritz Haber, who devised a process at BASF for synthesizing ammonia, which later allowed Germany to manufacture explosives and thus foil the Allied blockade during the First World War. A new fashion of conducting research appeared. As a result of the financial and human resources provided, research became more and more massive, diversified in its tendencies and targets, planned in its modes of organization, and collaborative through the increasingly close relations established among the different actors concerned with its results, particularly between those of university laboratories and those of business enterprises.

On the eve of the war, the model of vertical integration—marked by the development of intensive internal integration, by the cooperation of the academic world, and by a collective effort of research supported by the state—was beginning to take shape. Between 1914 and the 1960s, it continued to strengthen, and a movement of international concentration followed. The model of huge and planned integrated research in enterprises, in close contact with universities, became generalized. Cooperative research took off. Furthermore, institutions and scientific laboratories financed by the state developed policies of research and innovation directly connected to enterprises.

This process of concentration depended heavily on strategies of research and development within enterprises, since large firms possessed the means necessary for research efforts. In fact, the growth of research in the large European and American enterprises after 1910 was astonishing. The best example can be found in a report edited by the dye firm of Britain's Imperial Chemical Industry (ICI) in 1936: "Our main problem is the highly competitive character of research in organic chemistry. The confidence of I.G. Farben in the future of this sector is so strong that its effort of research is preponderant. That is demonstrated by the patents obtained by I.G. Farben, which represent three-quarters of those in the realm that interests the dye industry.... The people at I.G. Farben have constructed a veritable barrier of patents."[10]

Some statistics suffice to illustrate this expansion. In the United States, the number of researchers employed by enterprises climbed from 3,000 in 1921 to 46,000 in 1946. During the 1930s, US private businesses financed two-thirds of the spending on national research and development. The national average of newly founded laboratories was twenty-nine between 1899 and 1918, sixty between 1919 and 1936, and forty-three between 1937 and 1946. In 1925 I.G. Farben owned twenty-five large laboratories employing 3,700 persons. Its expenditures for research and development reached its peak at that time, representing 10 percent of its budget before falling to 6 percent in 1934–1935. R&D comprised 2.8 percent of the budget at ICI and 2.4 percent at DuPont. In France, this movement did not begin noticeably to expand prior to the 1930s, but it accelerated rapidly after the Second World War.

Between the two world wars, national and international oligopolies were constituted in most of the large sectors of major technology such as chemistry and electronics. Lifted to a global status by the multinationalism of enterprises and by international accords, they rested on shared knowledge among the territories where they flourished. Unsurprisingly, in the long run these global technological enterprises—automobiles, electronics, chemistry—had a tendency to stagnate. After 1945, the sole new entrants were large Japanese firms as well as others in electronics and computers. In the 1960s and 1970s, these sectors underwent the same process of rapid concentration as did those at the turn of the century.

This made possible creative strategies for diversification and new products. Wonders were performed at AT&T's Bell Laboratory. Fundamental research there resulted in a considerable number of major innovations in electronics, such as the transistor in 1947. Yet Bell's monopoly was unable to survive the liberalization of the telecommunications market that began in the 1970s. Moreover, several new innovative firms in this sector were created by researchers emerging from the laboratory itself. As it happened, the model of vertical integration became unsuited for a competitive environment, and its utility was increasingly contested after the 1970s. The direct connection between fundamental research and new products or procedures appeared more uncertain, if not improbable. The adventure of nylon produced by DuPont in the 1930s, after discoveries by Wallace Hume Carothers, was not repeated as often as one might imagine. It was necessary to return to a conception of research based on a perception of technological restraints and market demands. Enterprises thus entered into a realm of techno-science in which technology was no longer apart from science and where science was neither pure nor necessarily discursive.[11]

Enterprises were directly involved in the emergence in the twentieth century of what could be called mega-science and of big technologies. These extended the trend of gigantism of research programs before 1914, of which the perfect illustration was the synthesis of ammonia by BASF. In the inter-war period, programs of collective research grouped several enterprises, many of which were created within the framework of professional organizations. The Calico Printers' Association (CPA), a UK textile company founded in 1899, created a research group in 1926 that applied for 202 patents by 1949, 116 of them in chemistry, among which was the 1941 invention by John R. Whinfield and James T. Dickson of a polymer that was later developed into the synthetic fiber Terylene by the ICI.[12]

In the 1930s the large worldwide petroleum companies, led by Standard Oil, were searching for a refinement process in order to avoid payments to Eugène Houdry, a French engineer and naturalized American citizen, who had invented the first catalytic procedure. The Catalytic Research Associates (CRA) was a consortium of a thousand researchers who conceived the first process of continuous refinement. This type of research was developed after the Second World War both by individual firms like the CPA and by collective enterprises such as the CRA. A good example of the former model was the Institut de Recherche de la Sidérurgie (IRSID) in France, whose work in the 1980s allowed the French metal industry to recover. The latter model was illustrated by the development, especially in Japan and the United States, of strategic alliances among enterprises that flourished in the 1960s. The aim was to share the growing expense of research and to reduce the costs and delays of technology. These models also responded to the increasingly interdisciplinary character of technological and scientific research by implementing coherent policies of development and research. These various factors explain not only the cooperation among enterprises but also that between enterprises and public and university laboratories.

In the United States, direct collaboration between enterprises and universities developed in the inter-war period. The policy of the Massachusetts Institute of Technology (MIT) in the 1930s was particularly innovative in this regard, and it continued to thrive after the Second World War. Such programs were oriented at the same time toward fundamental as well as applied research, whereas the cooperation among business firms tended above all to generate new products. These programs grew to gigantic proportions during the Second World War and the Cold War. Confined to basic physics in the 1930s, they expanded after 1945 to include nuclear and space research, but also electronics and aeronautics. Some of them entailed such exorbitant expense that only states could finance them. Businesses nonetheless largely continued to participate in them and benefited from the knowledge that was engendered. Hence, these programs depended on close collaboration between university scientists and engineers from enterprises.

The Manhattan Project factory that produced the world's first atomic bomb was designed by DuPont's chemical engineers. Without their know-how, there would have been neither a factory nor a bomb to end the war in 1945.

Enterprises thus appeared as privileged sites for the convergence of knowledge united in a collective effort to exploit their capital in meeting market needs. Each of the participants belonged to a professional community for which the acquisition and construction of knowledge were unique. These particular traits must be appreciated in order to define the integration of various artisanal trades, engineers, and academic cultures.

Artisanal Trades and the Formalization of Knowledge

Artisanal knowledge was formed on the basis of apprenticeship, which led to a definition of tacit rules transmitted from generation to generation. Technological knowledge contained in the earliest factories of the nineteenth century could extend this artisanal know-how only by becoming sites for bringing different trades together. Technological change brought the progressive disappearance of some artisanal know-how while ensuring its constant renewal. Mechanization in particular provoked degradation by transforming versatile laborers into specialized workers who were reduced to executing certain simple, repetitive gestures. But this process alone does not account for the breakthrough of quantification, which also resulted from a continuous restructuring of trades marked by the persistence of some of them, the disappearance of many others, and the development of new skills in new technologies. Empirical knowledge generating fresh know-how was ceaselessly inaugurated and modified in the process of production. But in all trades, old and new, the more or less well-educated artisan had first of all to master his gesture, his body, and his muscles. As Yves Lequin has remarked, it was also "a matter of the eye ... when the glassblower or metalworker opened the oven; of the ear when listening to the noise of a machine or a construction site; of the nose or skin when forced to step away from the fire door of a blast furnace."[13] Above all, it was a matter of adapting tools to the task. Each trade was identified with one or several tools and depended on muscle memory. On the eve of the First World War, in an industry as symbolic as ironworking, tacit and empirical knowledge of this type was still predominant.

In the course of both the first and second industrial revolutions, new trades appeared that made use of new technological systems. In the first industrial revolution, it was engineers of locomotives and steam engines, puddlers, and machine operators who invented empirical know-how through their experience, ingenuity, and intelligence. Railroad engineers had to create a synthesis between their high level of manual know-how and a broad conception of the functioning machine that they drove or repaired. In the second industrial revolution, it was winders and casters in the mines, metalworkers in the aluminum industry, and electricians and machine operators in all industries who forged this synthesis. The transmission of these forms of knowledge was accomplished through apprenticeship to which instruction in specialized schools was added in some trades after 1850. Artisans and workers in workshops not only improved trade practices, gradually introducing innovations in trades and machines as well as manufacturing processes, but also played a major part in the more radical changes that led to the development of some new industries, such as bicycles and automobiles. The machine tool, automobile, and airplane industries were largely the result of these workshop innovations of qualified workers and artisans. Thus, the brake shoe was invented by a Michelin worker in the 1880s, and the first De Dion-Bouton autos were made by two Parisian artisans.

Much of this know-how that had been accumulated by workers and artisans could be transformed into formalized and codified knowledge. Manuals were edited and instruction offered in technical schools or specialized courses. Thereby the professional worker became qualified as a result of tacit knowledge gained through experience as well as codified knowledge. The mechanic and the electrician provided the best prototypes of this evolution. Machines and electrical installations required a thorough understanding of their function. But this formalization did not exclude knowledge acquired in the exercise of a trade; rather, that knowledge became the pedestal on which engineering science was constructed. In order to master and to develop engineering science, said Karl Marx, first "the veil of professional secrets" had to be lifted. The transformation of those secrets into formalized knowledge was done directly by an engineer who, after observing and analyzing artisanal procedures and, when necessary, employing his own learning, was able to construct a sort of protocol in which he sought to introduce a greater rationality inspired by the results of analysis and mechanization.

A good example of such a process is furnished by the history of glass manufacturing in the factories of Saint-Gobain at Chauny in the 1830s and 1840s. This industry, as Jean-Pierre Daviet has noted, comprised "in its traditional form a certain margin of play accorded to what might be called cleverness, flair, intuition, or the observation of empirical facts that the science of that time had absolutely no way to comprehend."[14] Quite early on, the directors of Saint-Gobain attempted to improve their production by applying a policy of innovation based on collaboration between the greatest chemists and professional mechanics of that era. They began by obtaining machines whose design was largely inspired by the gestures of workers in order "to rationalize existing empirical technology."[15] In the 1840s, they recruited graduates of the École Polytechnique, including the son of the renowned chemist and physicist

Louis Joseph Gay-Lussac, but none was able to achieve the necessary fusion of knowledge. In 1852 they engaged Hector Biver, a graduate of the École Centrale, who had previously worked for competing Belgian firms and had concrete experience in the thermochemistry of glass and pottery. Hence, "he did not need to receive instruction from a boss of the Paris markets to penetrate the mysteries of glass-making."[16] Already knowledgeable about some of those mysteries, he could learn the rest. Thus, barriers fell and Biver's success was assured. The tension that had arisen in the 1840s between workers and factory owners subsided. By consulting with operators, Biver was able to reach a synthesis between scientific and empirical knowledge. During the 1850s and 1860s, he dealt with a strong rise in demand and obtained a notable reduction of costs due to a series of innovations that improved current practices in pottery-making and in the regulation of ovens, which allowed a more methodical control of temperatures. With the aid of his father-in-law, Théophile Pelouze, and as a result of his own concrete understanding of the process, he was able to adapt given procedures to change. He thereby succeeded in combining two types of knowledge that, after having been in competition, now became complementary. He accomplished this by appropriating the know-how and skills of workers and rationalizing them while adopting certain technological artifacts created by science. The automobile, electronics, and aeronautics industries could not have developed without this mobilization of "accumulated knowledge and maneuvers derived from individual trades" in mechanics, to recall a statement by d'Alembert in the *Encyclopédie*. And in the 1840s the curriculum taught at the École Centrale in Paris perfectly illustrated the proximity between artisanal know-how and engineering science.

Engineers and Engineering

Artisanal trades and workshops thus contributed directly to the birth of an engineering culture that was increasingly competitive and that was taught in specialized schools. This is why some authors have defined it, contrary to the trade culture, as a school culture advanced by the creation of engineering academies and the development of scientific instruction with a strong technological component in the universities.

Education of Engineers

In 1856, during a celebration of the twentieth anniversary of the founding of a technical school in Hanover, one of its professors defined the mission of such institutions as being to "carry the blazing torch of science into the twisted corridors of industrial activity so that spirit may animate matter and rule the hands."[17]

This apt formulation, without being exact, does not entirely account for the reality of the actual content of technological knowledge practiced in workshops, academic circles, and laboratories. In the first half of the nineteenth century, the education of most engineers was the result of an apprenticeship in workshops or on construction sites. This was almost exclusively the case in Great Britain for both mechanical engineers and specialists in civil engineering. According to Robert A. Buchanan, the majority of engineers in the 1850s had acquired their know-how in the office of an active engineer "in a manner directly derived from experience acquired from practical education through apprenticeship."[18]

In France, the great Parisian mechanical industry, with men like Cail, Cavé, and Farcot, was created by autodidactic mechanics, simple workers, and artisans coming from machine workshops or clock-making.[19] Even in Germany, where technical education was more developed, autodidactic agencies were much more numerous in industrial enterprises than school-trained engineers until the 1860s. There, as in the United States, shop culture maintained its prestige and persisted as the basis of the mechanical sector until the end of the nineteenth century.

The apprenticeship of the engineer in a workshop or construction site rested on a dual initiative: first the culture of trades and then the shared experience of colleagues. Engineers were immersed in the world of trades, which in the nineteenth century underwent a radical transformation as each new technological branch created new practices or remodeled old ones. A young engineer also benefited from the experience of his colleagues, thanks to his participation in business activities. The system of apprenticeship did not consist solely of transmitting practical knowledge; through selection by management, it also established a hierarchical structure characterized by highly personal mentoring relationships. In all countries—and not only in the United States—the bond established between the young engineer and his mentor was an essential component of initiation that developed through experience acquired "in a bureau of design or on a worker's bench."[20]

The development of practical and formal instruction specifically devoted to the education of qualified technicians and engineers modified the inception and transmission of technological knowledge. Each nation constructed a unique system corresponding to its particular technological culture and educational tradition. Yet it is possible, according to Peter Lundgreen, to speak of a phenomenon of convergence that appeared in the 1860s and became fully manifest by the opening of the twentieth century.[21]

In Great Britain, the birthplace of the industrial revolution, it required more than a century to consolidate a complete and nearly coherent system of engineering education. Universities had the advantage, since three coexisting currents converged toward them. Chairs of engineering and evening courses were soon created at certain universities, especially in Scotland. Founded in

1842, the chair at Glasgow, considered to be among the most prestigious, was accorded in 1855 to William Rankine, the most acclaimed European specialist in engineering science. England followed the Scottish model belatedly. Chairs were not established at Cambridge until 1875 and at Oxford in 1907. A second current aimed at establishing university colleges such as University College London (1826), King's College London (1829), and Owens College in Manchester (1851). Initially oriented toward instruction in the sciences, they later took up engineering. Colleges of this type blossomed after 1870 in response to the needs of local industries: a dozen were established between 1871 and 1902, many of them in industrial cities like Leeds (1874), Sheffield (1879), and Birmingham (1880). More idiosyncratic were institutions outside of the university. The British navy and army created schools specifically for engineers such as the Royal School of Military Engineering, founded in 1812. The Royal College of Chemistry began operations in 1845 as a private foundation. Between 1853 and 1865, it was directed by the eminent German chemist August Wilhelm von Hofmann, and in 1851 it became integrated with the Royal School of Mines. In 1879 the Manchester Technical School and the Finsbury Technical College appeared simultaneously. Both taught mechanics, electronics, and applied arts. The Central Technical College was created in South Kensington in 1890, and the Imperial College of Science and Technology, modeled on the Technische Hochschule of Charlottenburg in Berlin, was established as a constituent college of the University of London in 1907.

The German system seemed to be the opposite of the English model. In Prussia and Saxony, using the French model, state engineering schools were created for mining, architecture (*Bauakademie*), and the army, but they were unable to secure positions as dominant as in France. In Prussia and several other states, starting in the 1820s, training schools appeared for technicians, qualified workers, and small entrepreneurs. In higher education, Prussia opened three commercial institutes, one in 1821 to teach mechanics and then electronics, another in 1850 for chemistry, and a third for naval technology. Other such institutions were established between 1820 and 1878, some of which gained a superior status. In 1879, the *Bauakademie* and the Institute of Mechanics merged to form the Technische Hochschule of Charlottenburg in Berlin, which became a model for others. Among the Technische Hochschulen dating from the 1820s, the first was Karlsruhe, which had a remarkable development, creating a school of engineering and a school of architecture in 1833, then a school of chemistry and a school of mechanics in 1850. This model caught on, and by the 1870s there were nine in all. These schools became the cradle of German engineering sciences and sites for the formation and affirmation of a genuine technological culture. Together they produced many more graduates than the twenty German universities. Not until 1899 was a doctorate of engineering conceived: the Technische Hochschulen were thereupon to become real universities of technology by establishing research laboratories in which instruction in physics and chemistry was introduced. In the sector of electronics, relationships with enterprises developed to such a degree that one could speak of an "industrial-scientific complex." Commercial institutes were transformed into technical schools, and schools of mechanics were created to recruit technicians at intermediate levels. The German model of education thus proved to be the most advanced by clearly asserting the autonomy of engineering science from university science.

In France, the corps of state engineers recruited from among the distinguished graduates of the École Polytechnique to form an elite group. Other graduates chose a civil career in railways or heavy industry. The three schools of Arts et Métiers that had been founded under the Napoleonic Empire, and which had suffered a decline during the Restoration, experienced a renewal due to a reform adopted in 1832. The curriculum now required a year of apprenticeship after which it combined workshop training with elementary education in mathematics, industrial design, and mechanics. These schools had been created by Bonaparte to train workers and workshop foremen destined for the mechanical and metallurgical industries, or, to quote the Emperor himself, "the non-commissioned officers of the industrial army."[22] Their actual role greatly exceeded that ambition. A large number of students rose rapidly in the industrial hierarchy and became real engineers or founders of prosperous enterprises in the leading sectors of the era.

The objective of the founders of the École Centrale, established in 1829 and made into a state school in 1857, was to offer scientific learning applicable to industry. If the program left ample room for pure science and retained an encyclopedic character, instruction there was nevertheless much closer to industrial realities than that of the École Polytechnique. Centrale became a seedbed of entrepreneurs and engineers who, confronted with state engineers, asserted themselves in civil engineering. In closer contact than others with the daily routine of production, they also had more concrete knowledge of industrial technology and hence importantly participated in the rise of industry during the second industrial revolution. Other schools, such as the École des Mines de Saint-Étienne, the École Centrale de Lyon, and the Institut Industriel du Nord, played an identical role. The last two, financially supported by local industries, were devoted to satisfying regional needs. But the generalist education of the students graduating from these schools put them in a difficult situation after 1880 because of the creation of twenty university institutes and several schools that specialized in new technologies—electricity and chemistry in particular. Among these specialized institutes were schools as prestigious as the École de Physique et de Chimie de Paris (1882) and the École Supérieure d'Électricité (1894). In addition, in certain universities, for example, Nancy and Grenoble, science professors began to take technological factors into account.

In the United States, the persistence of an education founded on apprenticeship coexisted with a more modest role played by state engineers in economic organization, thus favoring the early development of private institutions established to meet the need for training engineers. In 1824 the real-estate magnate Stephen Van Rensselaer founded, with Amos Eaton, a school that in 1849 became Rensselaer Polytechnic Institute. It was intended "to educate sons and daughters of farmers and mechanics" by applying science to "arts and manufacturing."[23] Gradually, the curriculum concentrated on engineering and technology, and a diploma in civil engineering was introduced in 1835. The reform of 1849 approximated the model of the École Centrale and the Technische Hochschulen. This institute, under the guidance of Benjamin Franklin Greene, became the first civilian center of higher education in engineering science in the United States. In 1846, William B. Rogers, author of a treatise on the durability of materials, proposed the formation of a technical school on the French model. This project resulted in 1861 in the creation of MIT, which offered courses in civil engineering, mechanics, mining, "practical chemistry," and, as of 1882, electronics. In 1870 the Stevens Institute of Technology opened at Hoboken, New Jersey, and another great engineer, Robert Henry Thurston, the first president of the American Society of Mechanical Engineers and a peerless leader in the technology of steam, became in 1871 its first professor of mechanical engineering. His goal was "to create a course of mechanical engineering that could serve as a model for purely technical schools with a mixture of general culture."[24]

In fact, the Morrill Act of 1862 put into place a system of subsidies awarded to states that opened schools teaching mechanical arts in order to encourage a synthesis between the humanities and practical education within the industrial population. After 1870, these schools abounded in America. The great scientific universities founded after 1840 entered into engineering studies only rather timidly starting in the 1860s. The first actual university school of engineering was created in 1867 by Dartmouth College, almost a century after its founding in 1769. Columbia University in New York opened a school of mines in 1864, which gradually expanded its instruction of the technological disciplines, including chemistry, in 1905. In 1885 Thurston took direction of Sibley College, which in 1870 had been founded as a division of Cornell University. His nomination symbolized the recognition by American universities of engineering as an academic discipline. And in 1879 a school of mechanics was established at Purdue University, followed by a school of mechanical engineering in 1882.

Thus, before the First World War, the great industrial countries had laid the foundation for teaching an engineering science that was coherent and capable of adapting to the evolution of technology. Altogether, scientific education was not sacrificed to a curriculum more oriented toward the ever evolving technology of production.

Engineering Knowledge

The culture of schools was not rigid. Rather, it was oriented toward a constant search for novelty and performance, even though the primary mission of an engineer is to make the current system function. The activity of engineers creates knowledge by itself, but unlike that of intellectuals this is not a major preoccupation for most of them. Above all, and often urgently, the engineer has to design, construct, and deploy technological systems for particular needs. To fulfill this mission, he must first mobilize the knowledge at his command. If it does not permit him to bring a coherent and complete response to questions posed by the task, he must find another means to solve them. In reality, the daily experience of workshops, factories, and technological networks is full of incidents and accidents due to more or less serious malfunctions of a technology. This constant technological uncertainty is particularly intense during the first stages of implementing a new technology.

The experiences of the first builders and users of steam engines for locomotives and propeller airplanes were painful. One of Edison's assistants recounted that after installing the first electrical lighting circuits from the New York center in 1882, it was considered a feat to keep the dynamo running during an entire day without an incident requiring a break for repairs. Such malfunctions revealed a complete inadequacy of parts or systems imagined by their inventors, since the gap between actual possibilities and the ambition for new systems is always considerable. Of course, it is feasible to make adjustments in the face of these situations. But the only solution is recourse to new combinations based on existing knowledge not yet exploited by the enterprise, be it by creating new strategies or new practices. Like the previously described process at the Saint-Gobain factory, the best way to do so in the nineteenth century was by improving the techniques of workers and artisans through formalization.

In the modern era, engineers took another path: they constructed formalized empirical knowledge obtained through their activities; they took advantage of scientific knowledge acquired in schools; they undertook experimental research inspired by intellectuals; and, as expressed by the Austrian engineer Ferdinand Redtenbacher, they sought "to base engineering science on solid rules," as he himself had done by publishing in 1855 a book entitled *Die Gesetze des Lokomotiv-Baues* (*The Laws of Locomotive Construction*). This means of forming new knowledge was the principal motor of industrialization in the nineteenth century through the creation of a specific domain of scientific knowledge called engineering.[25]

A study of the curriculum at the École Centrale in the 1840s is a good introduction to the specificity of engineering knowledge in the early nineteenth century. Directors of the school clearly distinguished between this knowledge and scientific learning in general. The objective of instruction was "to inspire in students a spirit of invention, so necessary in the practice of engineering, while directing them toward useful pursuits."[26]

The most important site for constructing knowledge in engineering science was the enterprise. Engineers worked there as employees, directors, or advisers. Since the end of the eighteenth century, consulting firms directed by independent engineers were created and played an essential role in this effort. Such offices in England, created by men like John Smeaton, John Rennie, and Thomas Telford, devoted themselves to the conception and construction of transportation and urban networks—that is, civil engineering—and to the construction of hydrological installations and industrial plants. They were also prolific inventors.

In France these offices developed from the beginning of the nineteenth century as a response to the need for expertise created by the construction of railways and by industrialization. One of the most celebrated was founded by Eugène Flachat in 1833. It assembled elite students from the École Centrale and became, as Flachat said, a veritable "school of applied science." The young engineers he recruited cut their teeth there before undertaking brilliant careers in railways or industry. In the face of the dominance of state engineers, Flachat demonstrated the necessity of gathering the particular skills of civil engineers, who were closer to the reality of industrial technology and were more capable of finding appropriate solutions to problems encountered in its application and improvement.[27] The group that he formed became "a porous membrane between the theoretical aptitude of a university graduate and the technical mastery of an engineer."[28] The vast field of competences covered by his office required very diverse expertise, extending from the installation of furnaces and forges to the construction of docks, from railway projects to the paper and petroleum industries. Flachat alluded to the fieldwork of his engineers as follows: "At the site or factory to be established, it is necessary to shape, to mold, to melt, and to adjust the pieces of cast iron; to forge and fashion the pieces of iron; to make flat surfaces, the outline of structures, the masonry, the framework, the roof, all the fixtures.... For that, the engineer has only local resources of materials and workers.... Sometimes he even needs to create them."[29] Thus, consulting firms were places for the apprenticeship of engineers, but also, and especially, they were sites for contact between formalized knowledge acquired in schools and the tacit knowledge of workers and artisans. As such, they were also special places for innovation. They did not cease to play their various roles, particularly in periods of emerging new technologies.

A Global Model of Constructing Engineering Knowledge

Placed at the crossroads of concrete experience and theoretical learning, of tacit and formalized knowledge, of engineering science and pure science, the engineer realized in his daily activity a cognitive synthesis that Walter Vincenti has analyzed through a study of aeronautical design between the two world wars.[30] He has constructed an interactive model applicable to the ensemble of engineering practices that generate technology, and he has established a relationship among six categories of knowledge utilized by engineers: basic concepts, criteria and specifications, theoretical tools, quantitative data, practical considerations, and design instrumentalities. Vincenti has also identified seven categories of activity that generate knowledge: transfers from science, invention, theoretical engineering research, experimental engineering research, design, production, experimentation, and testing conducted during the operations of a system.

These different forms of knowledge and activity proceed in the following manner. At the outset there is a definition of fundamental concepts resulting from theoretical research and experimentation confirmed by tests and eventually by inventions. Then there is an apprenticeship by engineers to undertake new research, both theoretical and experimental, in order to achieve a transfer of knowledge from science and to conduct new tests to verify or improve models and procedures. Thus, a solid technological base is set into place that includes criteria and specifications, theoretical tools, and quantitative data. This new knowledge is examined through production, usage, and further testing. The "practical considerations" that emerge from observation and careful exploration of the productive process create fresh knowledge and may suggest new solutions or orientations, similar to the creativity of engineers who depend on individual skill. "[E]ngineers," says Vincenti, "work to embody requirements from practice ... into as concrete and definite a technical formulation as possible"; furthermore, "engineering devices are by definition made to be used, and feedback of knowledge from use to design is essential."[31]

To place these various concepts and stages into historical perspective, one must take into account situations in which solely practical and empirical know-how, most often derived from trade cultures, is available and in which the basic concepts of a fledgling technology are imprecise and blurred. Such situations are precisely those that characterize the emergence and early development of new technologies. On this basis it is possible to imagine a historical scenario by adopting Vincenti's concepts. To respond to technical uncertainties, empirical knowledge is first constructed and then formalized and improved through observation and analysis of practices, results of testing, and experiments. An initial theoretical form thus appears. In a second phase, theoretical and experimental research assumes a dominant character. This may produce a definition of basic concepts and a formulation of rules, standards, and laws expressed in mathematical language, all of which are subject to questioning at any moment. Hence, certain concepts or theoretical tools are pure products of practice and experimental research. Other concepts, such as *force vive* (live force), mass, and electrical current, are scientific in origin. Transfers between these two types of knowledge occur in both directions. If many technological impasses can be broken only by advances in scientific knowledge, to a great extent the latter evolves under the constant pressure of challenges by technology. In numerous instances, engineering science corrects errors in scientific trajectories and blocks the false leads of intellectuals.

The Development of Engineering Science: Three Examples

Several exemplary trajectories show how engineering science developed in an autonomous manner by creating a corpus of coherent knowledge that allowed for a definition of precise rules and thereby clarified its relationship with academic science. What follows is an analysis of, first, hydrology and the technology of turbines; second, the means by which the science of durability of materials was created in the United States; and, lastly, the emergence of chemical engineering as an autonomous discipline.

Hydrology and Turbines

In the first half of the nineteenth century, one of the earliest examples of a distinction between theory and practice was the relationship between theoretical and applied mechanics. At first glance, there seems to have been a total separation between the mechanics of Joseph Louis de Lagrange, adumbrated in his "entirely mathematical" treatise of 1788, and the reality of industrial mechanics, as well as between the latter and the geometric theses of Gaspard Monge and Jean Victor Poncelet. However, these two scientific disciplines, by opening the way to a mathematical and geometric conception of mechanical phenomena, became the principal source of inspiration for the founders of applied mechanics. As Edward W. Stevens writes, it is clear that without mathematics the technological advances of the nineteenth century would have been thwarted. Stevens adds: "The nineteenth century surely was a great age of mechanization, and the technique of invention itself was rationalized by perspective geometry," without which contemporary mechanics would never have seen the light of day.[32] A famous remark of Albert Einstein aptly describes this situation: "As far as the laws of mathematics refer to reality, they are not certain; and as far as they are certain, they do not refer to reality."[33] Obviously, that by no means suggests that mathematical laws are unnecessary for comprehending reality. The often repeated contentions of engineers that the laws of physics do not apply to unduly complex technological realities always end by being disproved: physical scientists are obliged by observation of nature and of technology either to abandon them because they are erroneous or to fathom their meaning by using mathematical or geometrical tools.

At the beginning of the nineteenth century, a small group of French engineers from the École Polytechnique founded the science of applied mechanics. Some were oriented toward schools of military engineering and artillery, others toward the École des Ponts et Chaussées. While conducting experiments in mechanics, they did so in a much more general and systematic fashion than Anglo-Saxon engineers. For them, experience was not an open door to invention. It was an instrument to measure and verify the basis of theory.[34]

The history of hydrology illuminates the opposition between Poncelet's conception of the rather tight relationship between theory and practice and that of engineers like Benoît Fourneyron, who used experimentation independent from theory as a possible source of invention. It also illustrates the eventual reconciliation of the two approaches in a synthesis that confirmed Poncelet's method. At the beginning of this evolution there were the ideas elaborated by Claude-Louis Marie Henri Navier in his critical edition in 1819 of a treatise on hydrological architecture by Bernard Forest de Bélidor, published in four volumes from 1737 to 1753. Navier offered a synthetic vision of fluid mechanics. He thereby opened a debate that found its theoretical formulation in the work of a young British intellectual, George G. Stokes, who studied at Cambridge during a time when the physicists of that university had recently developed teaching infused with French mathematical physics. Although he was poorly informed about problems posed by hydrological engineering, Stokes introduced the concept of viscosity to account for the discrepancy between the theoretical and actual flow of water. By synthesizing the works and experiments of French and English engineers and intellectuals of that era, he succeeded in formulating what is known as the Navier-Stokes equation. According to Olivier Darrigol, it is "the only hydrodynamic equation that is compatible with the local isotropy and a linear dependence between stress and distortion rate."[35] Yet the Navier-Stokes equation was of no immediate use to hydrological engineers. If in principle it contained all the mechanical processes of fluids, it could not be applied except in a very small number of instances related to technology.

The vertical wheel with curved paddles described by Poncelet in 1825 was inspired by Navier's book. It was "the first example of machines conceived from theoretical reasoning."[36] Poncelet, who wanted to create an efficient industrial tool, calculated the curvature of paddles, the volume of water, and the size of wheels in order to obtain a maximum effect, and he curved paddle wheels to match the speed of water. Then he moved to experimentation "to utilize his propositions as a method of invention."[37] Poncelet's invention of the waterwheel was thus improved and its performance increased. But when the results of his research were published around 1870, his wheel was already obsolete, and its use was limited to small waterfalls.

However, Poncelet had designed another wheel in 1823, this one horizontal. It was long considered the precursor of a turbine invented by James Francis, which had great success. Poncelet's original idea made the rounds of Europe and the United States during the 1830s and 1840s. But testing of Poncelet's conception by different builders in the 1840s arrived at mediocre results, as the French physicist General Arthur Jules Morin confirmed. Poncelet's setbacks showed that a theory of machines furnished neither the calculations nor the knowledge "to channel the turbulent movement of water into a wheel or turbine," and that "only experimentation, coupled with a certain mechanical intuition, could achieve fundamental technological innovations."[38] Such was the procedure followed by Fourneyron in France, who was the true inventor of the turbine, as well as Americans like Francis and Lester Allan Pelton.

Fourneyron's turbine was derived from the horizontal wheel designed by the Swiss mathematician and physicist Leonhard Euler in the eighteenth century and improved by Claude Burdin, one of Fourneyron's professors at the École des Mines de Saint-Étienne. Becoming an engineer in an ironworks in the Department of the Doubs, Fourneyron wanted to improve the waterwheel of the factory in order to manufacture tin. He designed a wheel of which the propulsive force was furnished by water ejected by paddles or flat boards on their edges. Thus, he inserted at least a part of theory into practice and laid the groundwork for a treatment of hydrological machinery. His machine followed the general principle enunciated by French hydrologists and by Lazare Carnot, since it allowed the penetration of water without shock. To obtain these results, Fourneyron performed a series of experiments not only to confirm the relevance of solutions suggested by theory but to find modifications in practice. Other installations in the Jura Mountains enabled him to perfect his system. He took out a patent in 1832 and was rewarded with a prize from a national industrial society in 1833. Poncelet praised Fourneyron's machine but regretted that he had proposed no new theory. Fourneyron did so before the Academy of Sciences in 1838. By then his system was fully developed: the turbines reached 2,300 revolutions per minute with an efficiency of 80 percent.

In the United States, the construction of water mills remained up to the nineteenth century an affair of builders. The knowledge that they put into practice was purely empirical in origin, and it did not begin to be formalized before the end of the eighteenth century. The brothers Austin and Zebulon Parker were among the most prolific American inventors. They relied on a simplistic theoretical base in complete ignorance of French research, but they accorded "a great importance to visual and tactile observation of water flow," and throughout their lives they conducted a large number of experiments. In the year following the invention of the turbine by Fourneyron, the Parker brothers introduced a waterwheel using a guidance system. Actually, as Edwin T. Layton has pointed out, their invention "depended not simply on an accident, but even more importantly upon a conceptual framework and experimental approach which was an outgrowth of a scientific tradition developed by American millwrights.

This tradition was the natural outgrowth of the craft knowledge transmitted by oral tradition from master to apprentice."[39] The Parker brothers pursued their experiments and formalized their results, although their work was not widely diffused. The hour of engineers had arrived.

Encountering competition from mill constructors after 1830, engineers were moved to create much larger installations. Such was the case of James Bicheno Francis, who was representative of the first American engineers. The son of an English hydrological expert, in 1833 he immigrated to the United States, where, after having participated in railway construction, he was engaged by a group of financiers promoting a vast industrial complex featuring hydrological energy, the Locks and Canal Company, which was located in Lowell, Massachusetts. Francis was to work for this company for forty years, during which time he also became a consultant and was sought after for his expertise in industry, especially in the hydrological sector. Although his education remained purely empirical, he kept informed of results obtained by theoretical studies and experiments in Europe, particularly in France. He published nearly 200 articles in which he merged the two traditions—that of French engineers and that inherited from British and American mill designers.

It was in 1843, after the publication of an article by an American engineer, Ellwood Morris, that the first Fourneyron turbine imported into the United States came to the attention of Uriah Atherton Boyden. A mechanical engineer and a close friend of Francis, Boyden was working in Boston at the time. He improved the design of the Fourneyron turbine by placing a distributor on the first one installed at the Lowell company in 1845. After Fourneyron sold his rights to the Lowell enterprise, Francis used a purely experimental method between 1849 and 1855 to perfect a new turbine, adapted from Fourneyron's, which was well-suited to medium and small waterfalls. Water ran not only above the wheel but also below it. Although Boyden and Francis did not ignore French hydrological theories, they believed that those theories did not sufficiently reckon with the effects of friction, water level, and internal resistance, for which only experience could supply a solution. Thereafter, starting in the 1860s, builders utilized the experimental method, confronted as they were with a reality for which theory could not account, despite its great sophistication. In the United States, increasingly efficient machines were conceived by applying a research and development policy based primarily on a method of trial and error. Many testing laboratories were created in which engineers eagerly multiplied their experiments. For example, the James Leffel firm, which introduced a double turbine, tried out nearly 200 different models before choosing the most satisfactory one. Thus, a basically experimental technology was elaborated whose procedures were not a simple transposition of scientific theory to reality. A different path was chosen because the objective, unlike that targeted by scientists, was the improvement of technological performance.

The American technology of turbines evolved along three trajectories. First was the improvement of the Francis turbine, which came to replace French turbines in the early hydro-electrical installations. After 1900 this equipment was challenged by a new type of machine, the turbine invented by Lester Allan Pelton for the exploitation of gold mining in California. Pelton's turbine adopted a procedure that had been used for a long time in certain mountainous regions such as the Pyrenees. In this method, water was projected onto the base of wheels with buckets in the shape of hearts. Pelton conducted research to improve the performance of wheels, buckets, and nozzles. He obtained his first patent in 1880 and opened a business in 1888. Pelton's system, well adapted to high waterfalls, dominated this segment of the market after the 1920s. In 1924, after lengthy research, an Austrian engineer named Viktor Kaplan produced a propeller turbine suitable for smaller waterfalls. It utilized paddles that could be adjusted to the speed of rotation. This procedure incorporated the most recent advances of research in fluid mechanics prompted by the needs of aeronautics. In the 1920s and 1930s intellectuals such as Geoffrey Ingram Taylor and especially Ludwig Prandtl introduced the concept of "boundary layers," which provided engineers in the hydrological and aeronautical industries with instruments enabling them to orient their research advantageously. They thus originated fluid mechanics, both pure and applied, which spread throughout the twentieth century.[40]

The Durability of Materials in America

The creation of a science of the durability of materials in the United States provides another example of the evolution of engineering learning situated on the frontier between experimental method and theoretical knowledge. Between 1830 and 1860, several programs to test the durability of materials were undertaken. In 1830 an inquiry led by Alexander Dallas Bache was conducted by the Franklin Institute into the causes of explosions in steam boilers. Then, in the 1860s, experiments on steam engines and the durability of cannons were financed by the federal government. In 1838 a treatise by William Barton Rogers appeared in which mathematical theory was used for the first time.

In the previous year Dennis Hart Mahan, a professor of military science at West Point, published a manual influenced by Navier and based on his acquaintance with French theories. In 1847 Squire Whipple, a civil engineer, published an elementary treatise that ignored French writings and embodied an extremely primitive approach. These tentative essays were insufficient to establish a coherent doctrine allowing engineers to find an adaptable solution. Thurston's first two great laboratories of engineering at the Stevens Institute of Technology and then in 1885 at Cornell University brought the answer. After the Civil War, a generation of intellectuals, educators, and engineers like

Henry Turner Eddy appeared. They were able to develop rigorous methods of empirical research equal to those employed in physics. But while they were developing abstract theories, engineers used a language based on the description of instruments. While this difference was manifest in all aspects of the American physical sciences, it did not prevent exchanges between engineering science and pure science, despite the divergence, clearly perceived by industry, between the cultures of these two disciplines.

Chemical Engineering

Another illustration in the growth of engineering science before its integration into a global system of knowledge was the emergence of chemistry as an autonomous scientific discipline. This process began during the final years of the nineteenth century and followed different paths in Europe and the United States.

In the United States, a strictly pragmatic current of research and teaching began in the 1880s based on the work of George Edward Davis, the British chemist considered to be the founding father of chemical engineering. Davis proposed to inventory with great precision the slightest details of procedures in the chemical industry, to classify them, and to put them into order. All the processes and basic operations of chemical technology were analyzed, and in 1901 Davis published *A Handbook of Chemical Engineering*, a two-volume manual of more than a thousand pages. His method of analysis arrived at the definition of an empirical model generally applicable to all chemical industries without distinction. In the United States, this notion was further studied by Arthur Dehon Little, who helped to develop chemical engineering at MIT and whose accomplishment in this field was comparable to that of Frederick Winslow Taylor in the mechanical industry. Technological units such as distillation, filtration, mixing, grinding, and brewing, were defined and analyzed pragmatically in light of a phenomenological theory of nature. Thus was born a clearly distinct profession of chemical engineers for which a specific curriculum was created. The German approach was appreciably different. There the new discipline was rather the product of cooperation between chemists and mechanical engineers. Research laboratories introduced new methods and procedures, but a fusion between chemical knowledge and the engineering corps of chemists was not actually realized. Meanwhile, until the 1920s, the analysis of complex industrial operations remained purely descriptive in the absence of a genuinely theoretical approach.

Beginning in the 1930s, chemical engineering became integrated into the sciences of physics and chemistry and was transformed into a real procedural science. If phenomena were analyzed with concepts borrowed from chemistry, hydrology, and thermodynamics, the discipline nonetheless remained dependent on a culture of workshops as a major source of innovation. It rested on a persistent effort to adapt industrial practices to new knowledge obtained through the use of instruments and measurements. Hence, the attentive observation of phenomena still constituted one of the principal sources of this innovation. The role first played by DuPont's engineers in the manufacture of nylon and then of the atomic bomb illustrates the importance of chemical engineering in technological systems. For nylon, a synthesis between the results of scientific research and the culture of workshops was rather easily achieved, because both benefited from it. This was far more difficult in the case of the Manhattan Project factory in Hanford, Washington, which employed chemical engineers from DuPont and the intellectual designers of the bomb assembled at the University of Chicago. "These two groups," writes the historian Pap Ndiaye, "did not speak the same language. It was necessary to take the results of research and translate them into a design for equipment." This translation took the form of plans and blueprints completely alien to the scientific culture of intellectuals. Hence, the Manhattan Project offered engineers an occasion "to demonstrate their versatility, since they passed in a brief time from nylon … to nuclear power." Intellectuals felt frustration when faced with chemists. Yet the production of the atomic bomb, according to Ndiaye, was "the first encounter between two currents that had until then ignored each other: one a half-century of research in nuclear physics … the other a half-century of mass production in industrial chemistry that began at the outset of the twentieth century with the synthesis of ammonia and techniques of catalytic high-pressure chemistry, which reached its zenith with the production of nylon in the 1930s." This fusion of knowledge was gradually reached throughout the chemical industry, following a model closer to the German precedent than the American.[41]

Conclusion: The Development of Engineering Sciences

In sum, engineering sciences developed according to the scheme that follows. At the outset, there were many zones of uncertainty and problems without solutions. Accidents were frequent, since the unknowns were sources of many malfunctions. To deal with this instability, engineers needed first of all to analyze and comprehend the logic of artisanal know-how in order to rationalize and formalize it. They could then construct theoretical knowledge codified from their own experience with production and with organized experimental research utilizing scientific learning acquired in schools. They could establish a fruitful dialogue between "useful" knowledge and theory, between engineering and pure science, either because they possessed a dual competence or because they engaged in an ongoing dialogue with the world of intellectuals. Since the objective of engineering research was to introduce new products and procedures, it could not ignore advances in fundamental science, whose goal was to fathom nature and the material properties with which those products were made and the physical environment in which those procedures functioned.

Yet it is altogether legitimate to distinguish an engineering culture from that of intellectuals. A transition of these two cultures was gradually achieved at the interior of the technological and scientific disciplines. It also came about due to pressure from civil engineers. But the engineers and intellectuals who had graduated from French engineering schools now appear to have had the most coherent vision of the techno-scientific civilization that was germinating. They enunciated the supremacy of a theoretical approach founded on mathematical models without shrinking from the gap between reality and the perfection of models. In their eyes, this gap could only be temporary and would surely be overcome one day. From this vantage, the adoption of French mathematical science in the 1830s by British philosophy, both at Glasgow and Cambridge, and then by American schools constituted a point of departure for the rise of different disciplines constructed "to comprehend a range of phenomena" whose complexity was first exposed by such exclusively technological objects as the horizontal mill wheel and the steam engine.[42] If the pragmatic approach of English engineers provided solutions for some immediate problems, the techno-scientific orientation of French state engineers, to which British intellectuals rallied, opened the way to the formation of contemporary physics.

Science, the Universities, and the State

The development of universities in nineteenth-century Europe inspired a radical transformation in the construction of scientific and technological knowledge. Intellectuals became professionals, pursuing secure careers and enjoying much prestige and a great autonomy from the rest of society. They specialized in increasingly delimited and narrow disciplines. To cite but one example, in 1967 those engaged in solid-state physics saw their specialty divided into twenty-seven sub-specialties. These restricted and often unstable research networks were the principal agents of innovation and of social control over researchers. Each of them was organized hierarchically on the basis of individual evaluation procedures concerning the quality of knowledge produced, a process that depended entirely on peer judgment of results. This system rapidly proved to be efficacious as the number of public and private universities grew.

Several national models of university coexisted. The German model was the best adapted to the rapid evolution of knowledge. It rested on strong competition among the universities and on methods of teaching oriented toward research apprenticeship and explanation of knowledge, whereas the French model was centralized and more inclined toward oral transmission of knowledge. The Anglo-Saxon model combined these two approaches but was more strongly influenced by the German model. These different models tended to merge without losing their national or local specificity. Most often, universities located research within a consciously inherited tradition.

The academic world also organized networks of long-standing or newly created autonomous university centers, and relations among those centers intensified. Contacts of personnel multiplied, thanks to the development of postal communications and advances in land and sea transportation. The identity of disciplines increasingly coalesced around three poles: associations of specialists in a recognized discipline or groups that sought such recognition; congresses, meetings, or exhibitions organized by universities or such associations; and the publication of journals that promoted contacts as well as the discipline. The great majority of universities in the United States and Europe rejected a utilitarian approach to research and asserted the pre-eminence of fundamental science over applied science. The goal of research was knowledge of nature and that of researchers was recognition by peers. This ambition was a means to conquer prestige related to both learned and disinterested activity and to foster the independence of scientists vis-à-vis the state and society. Actual practice was in fact largely distanced from grand principles, except in a somewhat forced manner in military research. National identity, increasingly powerful in the nineteenth century, drew intellectuals into this path in war and peace. They were without much hesitation the inventors of poison gas during the First World War and of the atomic bomb in the Second World War.

Well before the 1950s, the academic world entered, often robustly, into the area of research called "applied," that is, research used to conceptualize products and to implement procedures. During this time, ties between business enterprises and intellectuals were formed, although these relationships were not always easy to manage. Researchers participated either as outside advisers to enterprises and research centers or directly in enterprise laboratories. However, the ethic of university research seemed contrary to that in enterprise or in centers of applied research, where the savant lacked the freedom to choose his field of research and where the evaluation of the results of research no longer depended solely on the judgment of peers, since also taken into account were the effects on industrial utility and consequently on profits. Although certain academics scarcely tolerated the contradictions between these two orientations, many others deliberately chose research oriented toward practical application, especially in sectors of advanced technology. Still others could reconcile the two and easily pursue parallel careers, all the more so because the boundaries between the two types of knowledge became increasingly vague.

Thus, before the 1950s, the problems posed by the conception and function of technology could not find solutions exclusively in the creation of purely scientific knowledge. Hence, the relationship between science and technology—if it had ever existed—ceased to be uniquely a simple linear liaison of pure science with industrial application and became above all an exchange between these two different approaches, complementary in their dynamic.

The history of the relationship between thermodynamics and the steam engine in the 1850s as well as between industrial dyes and organic chemistry in the 1860s and 1870s furnished two clear illustrations, although they were very different in their evolution. Starting in the 1850s, while the cooperation between engineers and enterprises continued, the collaboration between business interests and universities gradually developed, despite the avowal of most academics that they were interested in fundamental and not applied science. They usually served as advisers, which persisted even after the appearance of more direct research activities by intellectuals within enterprises. Academics were thereby transformed into scientific and technological advisers of enterprises and oriented their decisions accordingly. Intellectuals as eminent as Lord Kelvin and James Clark Maxwell became recognized industrial experts and participated directly in the administration of telegraphic networks. Likewise, more than one French chemist became closely involved in the chemical industry, although this cooperation was sometimes boisterous, as was the case of Henry Le Chatelier in his relations with the Lafarge enterprise.[43]

Also, well before the 1850s, "hybrid" institutions were created, financed by the state or by enterprises—most often by both—to conduct research on the frontier between them. Around a common project they assembled professors, functionaries, engineers, and enterprise researchers in the name of state-coordinated national research. In this regard, the American and German efforts were ground-breaking. It was initially by creating testing laboratories and by fixing standards that the state entered into industrial research. Such laboratories were often directed by highly qualified scientists, and many became laboratories for industrial research.

In France, testing and standardization, such as the adoption of an official meter, followed from old traditions. During the Restoration, the Artillery Committee, founded in 1792, created a mechanics workshop and then a chemistry laboratory where Gay-Lussac and Sainte-Claire Deville worked. The Conservatory of Arts and Trades played an important role up to the 1860s due to Morin, who devised scientific methods of testing, particularly in hydrology. Unfortunately, his work was not pursued, and it was not until 1900 that a laboratory worthy of the name was installed at the Conservatory. The lack of progress in this respect was considerable on the eve of the First World War, and France's position in international negotiations concerning standards for scientific and technological instruments remained weak as a result.

In England, the hostility of entrepreneurs to state interference in this domain was difficult to overcome. Many independent professional laboratories were created, but not until 1899 was a national physics laboratory founded, which by 1918 comprised eight departments employing 532 persons. This was not a simple laboratory of testing and standards but a true center of research whose ambition was "to place at the disposition of the nation the power of science and by all means to destroy the barriers between theory and practice."[44]

The German model best anticipated the later evolution of the field. Beginning in the 1870s, several German states created testing laboratories at the request of industrialists. At the federal level, a Mechanisch-Physikalisches Institut opened in 1887. Proposed by Werner von Siemens, it came to dominate the German electronics industry. At first it encountered very strong resistance, but Siemens had the means necessary to prevail. Since his goal was "to integrate science and technology," he foresaw an ensemble that "would be scientific, with one section developing scientific research and another making technological use of scientific knowledge." Siemens was persuaded that pure science could go "hand in hand" with technological applications. He largely financed the institute, although he believed that the imperial government also had a duty to support it in order to encourage this interaction. Without ever becoming the great national research center envisaged by Siemens, the institute played a major role in fixing the standards of different sciences and of the technologies associated with them, especially electronics. Thus, it served to "thwart French ambitions in this area," ambitions clearly recognized by Siemens during the first world electricity exhibition in Paris in August 1881, which was followed by the creation in France of an electronics testing laboratory.[45]

It was not until 1905 that a new imperial institute of biology was established for agriculture and forestry, nearly twenty-five years after the institute of Siemens. It was followed in 1911 by the Kaiser-Wilhelm-Institut, which was proposed by three leaders of German chemical science, Emil Fischer, Wilhelm Ostwald, and Walther Nernst, as a counterpart to the laboratory of physics. This project, too, encountered strong opposition from agriculture, metallurgy, and in part chemistry, among other sectors. Financed by the Prussian government rather than by the Reich, this institute was divided into three distinct scientific branches (pure chemistry, physical and electronic chemistry, and experimental therapeutics), to which a department specializing in metallurgy was attached during the First World War. Essentially financed by enterprises of heavy industry and chemistry, and also by banks, in 1914 the Kaiser-Wilhelm-Institut found itself isolated in an ambiguous position, located between the university and industry. Yet under the direction of Fritz Haber, it took part in chemical research during the war, after which the Weimar government created a national institute of chemistry. In 1948, the Kaiser-Wilhelm-Institut became the Max-Planck-Gesellschaft, devoted to the advancement of the sciences.

During the First World War, massive research programs financed and controlled by the state had been launched. This trend continued in the inter-war period and grew to gigantic proportions during the Second World War and the Cold War. Such programs were soon no longer limited to military objectives, thereby creating so-called mega-science. This expansion of research had begun in the final third of the nineteenth century in fundamental science as well as applied science. It resulted in a progressive transformation of experimental practices, in the adoption of increasingly complex testing procedures, and, finally, in the utilization of scientific instruments that became more and more costly due to their technological sophistication and enormous size.

The cyclotrons constructed at Berkeley by Ernst Orlando Lawrence between 1933 and the 1940s, for which he received a Nobel Prize in 1939, may be regarded as the symbol of a new scientific culture marked by the huge installations and expenses that were required by physics by the end of the 1950s. Lawrence's machine of 1933 weighed 80 tons, that of 1938, 225 tons. According to Dominique Pestre, Lawrence's goal was not so much to solve the problems of fundamental physics but "to push ever farther the technological limits, to invent more powerful machines capable of maximum performance, to be the first to master a new range of energy," and like all other physicists "to produce the nuclear missiles needed."[46] Lawrence seemed more interested in technological performance than in theoretical physics. During and after the war, his experiments became part of global physics and resulted in the creation of the European Organization for Nuclear Research (CERN). The tendency toward gigantic installations and programs has thus dominated physics, which among all disciplines was—and still remains—the greatest consumer of research funds.

The two world wars, especially the second, decisively furthered not only the growth of programs but also the direct and increasingly indispensable participation in research efforts by the state. Before then, its primary influence had concerned armament industries and large public works. Its activity in research was more to incentivize than to take part. It financed academies, universities, and engineering schools, orienting production by placing orders for goods. Then during the 1880s, hybrid institutions appeared that were financed simultaneously by the state, universities, and enterprises. In a spirit much like that of mercantilism, these institutions were devoted to the promotion of scientific disciplines, such as physics and chemistry, that were considered necessary for the development not only of the armaments industry but also for national technological industries. In the course of the two wars, many research centers and agencies were created that merged science into the war effort. The symbol of this military mega-science, which mobilized both scientific knowledge and technological know-how, was the Manhattan Project to construct the atomic bomb, an American undertaking that commanded human and financial means on an unprecedented scale.

Chapter 6

Steam Engines

The history of stationary steam engines shows how different systems of knowledge were combined to develop a complex technological object by adapting it to increasingly numerous uses. Despite the astonishing diversity of solutions proposed by its inventors and builders, several steps were required to combine and integrate their knowledge for an engine that appeared by the end of the nineteenth century to be a symbol of industrial civilization—although in fact, according to Jacques Payen, it was nothing more than the product of a "saturated" technology.[1]

To simplify, we may distinguish two channels in the construction of this technology: one that concerns the solutions adopted in the use of steam, and another that was purely mechanical to regulate the flow of steam. Uncertainties abounded about the transformation of heat into motor power until the 1850s, when thermodynamics became established as a recognized scientific discipline capable of promoting a coherent application of engineering. This delay explains the sometimes tentative and incomplete character of many engineers' attempts to work out innovations improving the use of heat and steam. By contrast, innovations of a purely mechanical nature rested on a much more secure basis of "a neatly articulated arrangement of fixed and moving parts whose improvement depended simply upon inventive ingenuity along mechanical lines."[2] A programmatic relationship formed between these two channels, the first oriented toward understanding and mastering the different stages of transforming heat into power by defining rules for operating under ideal conditions, and the second devoted to the conception and production of machines best adapted to these constraints.

Domination of Empirical Knowledge before 1850

The World of Mechanics and Engineers

Until the 1840s, the use of steam engines was mostly restricted to sectors where their irregularity of motion did not unduly disturb the functioning of machines and where the power necessary for their operation remained limited. Yet a slow process of improvement was developing—including in the United States, where their use was still marginal. As Louis C. Hunter has noted, in America before 1850 "improvements in engine design and building were evidently far less the result of conscious intent, whether directed to specific components or the machine as a whole, than the sum of many minor changes suggested by the practical experience of builders and users and related especially to problems of maintenance and repair." Progress was thus the fruit of "a process of collective trial and error."[3] According to Hunter, workers at every level of the sector had training and an environment drenched with mechanics, even if they lacked sophistication. Hence the advances realized in building engines were at the same time incremental, collective, and anonymous. This type of innovation was not absent in Europe, but there a prominent role was played by an elite of engineers and entrepreneurs capable of conceiving new systems, such as the successors of John Smeaton and James Watt in Great Britain, the most brilliant of whom were doubtless Richard Trevithick, Arthur Woolf, Matthew Murray, and William Murdock, or in France self-taught Parisian mechanics like François Cavé and Joseph Farcot or Alsatian entrepreneurs like Hieronymus Stehelin.

In 1800, radical new innovations appeared in Great Britain. In the tin mines of Cornwall, where the firm of Boulton & Watt had established a major market and installed large pumps to evacuate water from mines, two great projects of steam technology were developed: the compound engine and the use of high pressure. Cornwall's mines were extremely deep, and mining companies consumed large quantities of expensive coal imported from Wales. They were also the first to purchase Watt's engine. Hence they sought to avoid dependence on unduly demanding suppliers who demanded exorbitant prices for patents. After 1800, and especially starting in 1811, the mining companies of Cornwall freed themselves from Boulton & Watt and adopted a policy of systematic competition among various inventors that enjoyed great success. They created a monthly journal describing the technological features and the performance of each machine. This publication stirred a spirit of competition among manufacturers that led to collective innovation, with only a small number of inventions being patented. The framework of steam energy was thereby radically altered. It was in this context that Arthur Woolf, Richard Trevithick, and several others were able to test their ideas. This atmosphere of competition and rapid circulation of fresh knowledge was not confined to the mining regions of Cornwall, even if it was most progressive there. All of Europe was swept by a wave of creativity.

The Uses of Steam: The Valve

The deployment of valves became gradually widespread after 1800. It was, according to editors of the 1845 edition of Charles Laboulaye's dictionary, "the greatest improvement in steam engines since Watt."[4] The valve enabled steam to expand inside of cylinders within certain limits. Promoters of the valve proposed to cut off the supply of steam after its introduction into the cylinder, thereby allowing the steam to expand, which would save both steam and fuel. Opponents doubted real gains through the early closure of the valve and claimed that it was necessary to inject steam steadily throughout most of the piston's action. This question, which concerned not only conserving fuel but also the basic principles governing the functioning of engines, could only be resolved by a better understanding of those principles. Thus, the limits of a strictly empirical approach became increasingly apparent and were evident to everyone by the 1840s.

The use of valves had other advantages besides economizing fuel. They prevented machinery breakages caused by violent operations. The trend toward increased pressure—more than twice as much as during the first half of the nineteenth century—made valves ever more necessary. The desire to master mechanical valves and the distribution of steam mobilized mechanical engineers and constituted an essential branch of research on automation. The problems posed by these engines, if one believes William Ernest Dalby, "have fascinated successive generations of engineers" to such a point "that there are few mechanical engineers who have not attempted at one time or another to invent new machinery."[5]

The history of slide-valves, which replaced ordinary valves after 1800, is enlightening in this regard. The first known slide-valves were developed by two British engineers already mentioned, Murray and Murdock. After apprenticeship with a blacksmith, Murray worked in different factories, including a large linen firm in Leeds. In 1790 he took out a patent for an improvement of textile machines. In 1795, with other engineers, he started a company to manufacture steam engines, and in 1799 he invented a system for regulating the intensity of heat from a boiler by utilizing slide-valves. Meanwhile, he used a control system with the purpose of guiding the flow of steam, a method later employed for locomotives. As for Murdock, he was the son of a millwright who received training and initiation to engines in his father's enterprise. Hired in 1777 by Boulton & Watt, he was assigned to the construction and maintenance of machinery in the tin mines of Cornwall. He thus acquired a deep knowledge of engines and designed a slide-valve that, after undergoing various improvements, was often employed to equip mills and maritime engines. The numerous advantages of the slide-valve explain the speed of its spread to other countries. Yet it was to be eclipsed after 1850.

The Uses of Steam: High and Low Pressure

From the standpoint of mastering thermodynamics, the two most important novelties after the elapse of James Watt's patent in 1800 were the compound engine and high pressure, closely related. Both innovations were tested out before 1800, the first by Jonathan Hornblower starting in 1781, and the second by Richard Trevithick in 1796–1797. Hornblower belonged to a family dynasty of manufacturers who were among the first to build the Newcomen engine and dare to defy the domination of Boulton & Watt. Hoping to recycle steam, he invented a dual cylinder device, but did not obtain pressure sufficient to outdo the performance of Watt's engine.

Trevithick also belonged to a community of Welsh engineers. Son of a mining foreman in the coalfields of Cornwall, he was educated on the job before becoming an engineering counselor of the same district in 1792. Even before 1800 he scouted out the possibilities offered by high pressure that might allow him to avoid the use of a condenser and thereby circumvent Watt's patent. In the decade after 1800, Trevithick installed more than thirty of his engines in Cornwall. They served in mines as well as watermills and rolling mills. He greatly increased the efficiency of high pressure by improving boilers. His most lauded claim to fame was applying this principle to the world's first steam locomotives.

Arthur Woolf, who had worked with Hornblower, took up his idea of dual cylinders. He lived in London, where he was in charge of steam engines in a large brewery. Experienced in maintaining them, he reckoned that it was necessary to invent more powerful, compact, smaller, and consequently more fuel-efficient engines in order to meet the needs of an industry using more and more power. In 1803 he took out a patent for an engine with two unequal cylinders in which steam moved from the smaller to the larger. Later there were three or four cylinders, in which the pressure dropped at each stage but the motor power remained the same. Theoretically, this recycling of steam would allow great savings of fuel, but Woolf relied on an erroneous theory of steam expansion by positing that the temperature would increase less rapidly than the pressure. He consequently adopted a much-too-reduced size for the high-pressure cylinder. Resuming his experiments, he increased pressure and used materials of the finest quality to succeed in achieving the same performance level as Watt with half of the fuel. In 1811 he began to partner with Humphrey Edwards, a millwright, manufacturer, and exporter of machines to France, where they later settled. There Edwards invented an engine twice as efficient as Watt's. Woolf returned to Cornwall, where his engines enjoyed great success. But the steam pressure in his engines was too weak to operate a compound system economically. Hence compound engines were not introduced until 1839, when James Sims devised a better distribution of steam channels and pistons. Edwards introduced Woolf's high-pressure engines in France, where they were better received than in Great Britain; Edwards subsequently supplied several hundred of them to the French.

Engineers in Cornwall returned to the single cylinder engine invented by Trevithick. Two of them, Joel Lean and Samuel Grose, were important players in its evolution. The brilliant results attained by their machines gradually became known in Great Britain and on the Continent. First employed in the 1830s for pumping water in London, this type of engine was later found throughout England. Actually, the Cornwall engines were not well suited for industrial use because of their large size, and also because of their inability to produce the regular rotary motion demanded by the great majority of industries. Yet they were simpler than the Woolf engine. As William Fairbairn wrote in 1853: "The dual cylinder engine, or compound machine, in which high pressure is used ... seems to assure a considerable saving of fuel, but if one takes two similar engines with a similar steam pressure, as is not the case for the best single cylinder engines, it appears that the compound has no advantage over the single cylinder. To the contrary, there is a loss due to the original cost of the machine and its complexity compared to the other."[6] For this reason Fairbairn did not hesitate to recommend single cylinder engines.

In the United States, it was steam navigation on interior waterways that first attracted engineers to this new source of energy. Very few of them were interested in stationary steam engines. Two technological developments were simultaneous. One followed the Boulton & Watt tradition of low pressure, an inclination that was reinforced by significant immigration of English engineers who participated in installing these engines in the United States before settling there, such as the mechanic James Sallman and the architectural engineer Benjamin Henri Latrobe. But this tradition rapidly lost ground to the high-pressure machines developed by Robert Livingston Stevens for steamboats and Oliver Evans for stationary engines. Evans was not only the inventor of a high-pressure non-condensing engine; he was also principally responsible for its success and diffusion. One of twelve children born to a Delaware agricultural family, he was apprenticed as a mechanic in a wheel factory, where he acquired know-how applicable to the manufacturing of mills. He soon became a prolific inventor in the technology of millstones. In the 1780s and 1790s he developed a complete system of continuous production that aimed to eliminate manual labor during the entire process. This system of conveyor belts was widely adopted throughout the United States, making him a rich man. Settled in Philadelphia in 1792 as a millwright, he sought to realize a project conceived in his youth: to fully exploit the properties of steam expansion beyond the possibility it offered to create a vacuum by condensation. In 1802 he built a very simple, small high-pressure engine, fashioned by a carpenter and a blacksmith and equipped with a powerful boiler. A larger model was later constructed and patented in 1804. In 1805 Evans published a work entitled *The Young Steam Engineer's Guide*. The English engineer Thomas Tredgold remarked that it presented "a strange mixture of absurd speculations and an indistinct perception of truth," whereas another commentator in 1866 saw "a quite complete and generally correct summary of the properties of steam plus a lucid and practical account of the advantages of high-pressure engines."[7] In reality it was above all a polemic for the view that the steam engine could be useful in all sorts of industries. In 1807 Evans opened a large workshop in Philadelphia and in 1811 another in Pittsburgh. Called the Columbian engine, this machine marked the real takeoff of high-pressure engines, of which Evans became the indefatigable promoter because they were, in his words, "small, simple, cheap, and light," with greater operational flexibility than low-pressure engines. Thereby Evans cannily foresaw the development of steam engine technology, albeit without the ability to realize all of his dreams.[8]

Structures

At the beginning of the nineteenth century, the wooden-beam pump was everywhere dominant because it was well adapted to evacuating water out of mines, which was still the principal use of steam energy. Such was the case in France, where Woolf's high-pressure engines were commonly in use. But mills and industries demanded ever better results, which explains the development of rotary engines and high-speed engines without condensers, both of which adopted a horizontal design that did not by itself increase the efficiency of steam engines but gave them greater utility and simplified their construction. Engines were thus more compact, more accessible, and always ready to start. The horizontal engine became widespread in the United States in the 1840s. After the international expositions of London in 1851 and Paris in 1855 featured their use, horizontal engines were deployed in Europe on railroads and in factories.

Through their empirical experience, engineers and mechanics thus succeeded in transforming the steam engine and adapting it to needs created by the variety of its uses. They constructed indispensable "useful" knowledge and established rules necessary for its production and exploitation. Many were involved only briefly, but others remained in the new field and shaped a ubiquitous corpus of recognized accessible learning. This evolution prefigured the process described by Walter G. Vincenti for the aeronautical industry of the twentieth century, in which certain rules were established "to embody requirements from practice ... into as concrete and definite a technical formulation as possible," and in which engineers were able to create new theoretical and quantitative knowledge in technology by using their own "judgmental skills."[9] Persistent research encountered several barriers. The fear of accidents hindered the adoption of high pressure, and the absence of a true science of energy, capable of describing the transformation of heat into force or explaining the loss of power during a machine's cycle, left free range for erroneous notions.[10]

At the outset of the 1850s, all engineers clearly recognized that the potentialities of steam engines were underutilized. Nonetheless, the avenues engineers and mechanics had opened since 1800 were not at an impasse. The intellectuals who were to construct the science of thermodynamics after 1850 would follow in the footsteps of their predecessors.

The Birth of Thermodynamics

Until the mid-nineteenth century, the understanding of steam engines and the sciences treating heat was not really a united body of knowledge. The steam engine brought many concrete results and facts to these disciplines, but intellectuals remained unable to explain the considerable gap between theoretical and actual performance of the machine, or to suggest effective means to improve it. A major turning point occurred during the 1840s and 1850s, a period marked by a brisk acceleration of theoretical controversy and experimental research that led to the birth of a science of energy founded on the laws of thermodynamics and the principle of the conservation of energy. The industrial milieu and the University of Glasgow—composed of professors like William Thomson, engineers like Lewis Gordon and William Rankine, and entrepreneurs like John Elder, all of whom maintained close personal relations—played major roles in this process of collective learning. This cluster of knowledge was integrated into a European network of information exchange that included all those intellectuals and engineers concerned with both the theory of heat and the proper functioning of steam engines, whether in Paris, Berlin, Mulhouse, or Manchester, to mention only a few of the many sites of innovation. These exchanges created an atmosphere of technological and scientific rivalry that proved fruitful for both.

Classical Thermodynamics

William Thomson's publication of a series of articles on "the dynamic theory of heat" between 1851 and 1856 was contemporary with the presentation in 1850 of a thesis at the Berlin Academy entitled *Über die bewegende Kraft der Wärme* (*On the Impelling Force of Heat*). These two events marked the beginning of a thermodynamic science founded on the conservation of energy. Rankine's work *Manual of the Steam Engine and Other Prime Movers*, which put the contents of this new scientific knowledge at the disposal of engineers, appeared in 1859 and was published in seventeen editions by 1908. This achievement was the result of a dialogue, often in the form of a controversy, that took place between physicists and natural philosophers on one side and those charged with building and operating machines on the other.

Four stages in this process of collective research may be distinguished:

1. The publication in 1824 of Sadi Carnot's *Réflexions* and their reformulation in 1834 by Émile Clapeyron.
2. The diffusion of Sadi Carnot's ideas during the 1840s and their confrontation with the results of experiments by several other intellectuals such as Henri Victor Regnault and especially James Prescott Joule, a Manchester brewer whose work, greeted at first with skepticism, represented one of the major breakthroughs in the history of nineteenth-century science.
3. The controversy that agitated European intellectuals around 1850, which was marked by a constructive dialogue between Joule and Thomson, then between Rankine and Thomson, and by a more polemical clash between Rudolf Clausius and Thomson.
4. The synthesis of thermodynamic theory and Sadi Carnot's model, achieved by Thomson himself.

Sadi Carnot and Clapeyron

The study of heat and its relationship to motor force is a very old tradition of European science. French mathematical physics first entered this path toward the end of the eighteenth century by adopting an approach brilliantly presented in a work published in 1822 by Joseph Fourier, *Théorie analytique de la chaleur* (*The Analytic Theory of Heat*). Although hardly understood at the time, it marked an essential step in the development of mathematical physics and the perception of caloric phenomena. In 1824 Sadi Carnot, then a young French mining engineer, published *Réflexions sur la puissance motrice du feu* (*Reflections on the Motor Power of Heat*), continuing this research and opening new avenues. This work applied theory to problems posed by the functioning of steam engines, particularly their weak performance. Even if this book was, as

Robert Fox has observed, "only one work among many others" devoted to the study of heat and motor power, such as those of Henri Navier and Alexis Petit, it presented a convincing theoretical summary of many facts relating to engines, derived from experiments conducted in physics laboratories and elsewhere for half a century.[11]

Distribution of Sadi Carnot's book remained very limited for nearly twenty years. Its rediscovery around 1840 was essential to the formation of thermodynamics. Sadi Carnot himself remarked that the *Réflexions* were intended as a response to questions posed by the functioning of machines using heat, especially steam engines. These questions had either a general character—such as whether the motor power of fire was limited, the temperature of heat sources needed to be raised, or loss of heat was avoidable—or a more precise character, asking about, for instance, the use of the high pressure recently introduced by Richard Trevithick and Arthur Woolf, or pressure's influence on the productivity of machines, making them more efficient and reducing their fuel consumption. Sadi Carnot could supply answers to such questions only by attempting to construct a general theory of heat. As Muriel Guedj has indicated, Sadi Carnot organized "his thought as a progression from practice to theory" through an intense effort of formalization.[12]

The spread of Sadi Carnot's ideas was slowed by his death in 1832. Émile Clapeyron, who was also a mining engineer, published a paper in 1834 in which he offered a clear analysis based on mathematics, drawn from the writings of Edme Mariotte and Louis Joseph Gay-Lussac, and displaying a graphic presentation. For this "diagram," inspired by a secret drawing by James Watt, Clapeyron used the coordinates of volume and pressure to provide an especially legible sketch of Sadi Carnot's ideas. This text thus became an effective instrument for disseminating them among both scientists and engineers.

It should be noted that Clapeyron subsequently became one of the most prestigious French railway engineers. It was he who, in France, propagated English methods of producing steam locomotives, from which an indigenous French technology gradually emerged. For fifteen years after 1844 he taught a course on steam engines at the École des Ponts et Chaussées. His 1834 work played a major role in the resuscitation of Sadi Carnot's thesis, and it was through him that the brothers William and James Thomson discovered Sadi Carnot in 1844. Clapeyron's mathematical formulas inspired those elaborated by Rudolf Clausius in 1850. In short, Clapeyron was more than a simple conduit of Sadi Carnot's ideas: he was a perfectly typical model of French engineering, both immersed in a scientific culture and able to explain a theory of heat that better corresponded to the preoccupations of engineers than did the theories of mathematical mechanics. He was also capable of incorporating his knowledge into a conception of technical objects that he devised and put to use.

Regnault and Joule

Two major changes occurred in European science during the 1840s. One was the penetration of Sadi Carnot's ideas into Great Britain, especially Scotland, and into Germany, thereby supplying a paradigmatic concept for the debate about the relationship between heat and motor force. The other was the increasingly precise experimental research into the measurements and formulations of resulting problems, particularly in Britain and France.

Several intellectuals undertook experiments to measure chemical and physical phenomena with precision in both the realm of electricity and the study of gas and steam. One of the most renowned laboratories was that of Henri Victor Regnault, a mining engineer, disciple of Gay-Lussac, and professor at the Collège de France. William Thomson visited France in 1845 expressly to learn about Regnault's methods of experimentation and measurement. There he acquired, as he later said, "a flawless technique, a love of precision in all things, and the most important virtue of experimentation: patience."[13] The French government, aiming to improve steam engines, tasked Regnault with measuring the loss of heat by diminishing gas or steam. Practice with laboratory procedures acquired from Regnault enabled Thomson to be, as he put it, something other than an "x and y man," and to become an expert in "mechanical operations."[14]

One cannot exaggerate the subsequent importance of this experience, which was obvious in the way Thomson organized his own laboratory at the University of Glasgow as well as in his theoretical reflections. The measures he adopted played a major role in Thomson's endeavors as well as those of Rankine and Clausius. Their models became directly involved in the dynamic process of constructing the knowledge that led to the birth of a science of energy.

Meanwhile, at Manchester in the 1840s, James Prescott Joule, scion of a brewing dynasty, pursued research on the release of heat by an electrical current. In 1843 he offered an initial evaluation of the mechanical equivalent of a calorie. Little persuaded, the scientific community demanded that he furnish further experimental proofs. He drove his experiments to the limit and succeeded in producing increasingly precise and convincing results by using, among other things, an appliance called a calorimeter. Thereby he was able to demonstrate that the totality of energy spent during compression was transformed into heat. Joule's demonstration of the equivalence of heat and power constituted a major advance in the history of physics in the nineteenth century. It lay at the origin of the definition of the first law of thermodynamics: that heat may be transformed into mechanical power and vice versa. According to Joule, mechanical force resulted from the consumption of heat and not its simple transfer, as Sadi Carnot had supposed.

Thomson, Clausius, Rankine

To understand the role played by William Thomson in the advent of thermodynamics, one must take into account the family and social milieu in which he grew up. His father, a professor of mathematics at the University of Glasgow, watched over his sons' scientific education. William was named professor at the same university in 1846, at the age of twenty-two. A genuine collaboration was established between him and his brother James, a professional engineer, in the 1840s. Later, in 1873, James was appointed professor of engineering in Glasgow. His family circumstances thus placed William Thomson at the crossroads of three traditions: Scottish natural philosophy, French mathematical physics embodied by Joseph Fourier, and general mechanics associated with the use of engines. Essential to this combination was Thomson's familiarity with the concrete problems posed by the invention and utilization of steam machines, evident in his relations with Lewis Gordon, the predecessor of Rankine and James Thomson at the University of Glasgow's chair of engineering.

Rankine was the son of a civil engineer. After studying natural philosophy at the University of Edinburgh, he first joined his father, who became director of a railway company, and then worked with other civil engineers such as John B. MacNeill. In the 1840s Rankine himself became a professional engineer specializing in the development of physical science in heating and lighting. He was named as Gordon's successor as professor of engineering at Glasgow in 1855. The observation that "the ideas of William Thomson on both heat and electricity in the 1840s expressed the practical reality of steam engines" can be extended into the 1850s.[15] Thomson's relationships with Regnault, James Thomson, and Gordon in the 1840s permitted him to come into contact with these realities. But in the decade thereafter, his contacts with Rankine became determinant.

In 1850 Rankine presented a paper on the mechanical action of heat and steam, adopting the idea of mutual convertibility of power and heat. Hoping to construct a theory based on the notion of a molecular vortex, he rejected the concept of calories because, he thought, heat was not a substance but a motion. He was the first to give public recognition to Joule's ideas, despite challenging his statistics. He was also the first to interpret Sadi Carnot in terms of conserving a *"force vive."* This was a decisive step toward reconciling Joule's theory with that of Sadi Carnot. Actually, Rankine's view was not solely theoretical, since he participated actively in research being conducted by several Glasgow shipbuilders, then leaders of British naval construction—especially one of them, John Elder, who was attempting to improve the performance of marine compound motors. In the 1850s, in close cooperation with Rankine, he built this type of more efficient motor. As Rankine indicated in 1871, the improvement of these machines "could only be realized ... by an engineer who studied and understood the principles of this path-breaking science of thermodynamics."[16] Elder had comprehended that, in order to take full advantage of a compound engine, it was not necessary to increase the energy produced by a given quantity of steam but to prevent losses of steam that reduced its force.

In the early 1850s a synthesis of Sadi Carnot's ideas as expressed in his *Réflexions* and Joule's theory of heat and power equivalence was achieved nearly simultaneously by Rudolf Clausius in Germany and William Thomson in Scotland. (Incidentally, they later engaged in a polemic about who had been the first to do so.) In 1850 Clausius advanced a thesis resting on the assumption that it was possible to completely transfer power into heat, whereas for the inverse that transfer could only be partial, which Carnot had already foreseen in his notes. Clausius concluded that the idea of the material nature of heat in a caloric form should be abandoned: "Heat is not a material but consists of a movement of bodily molecules."[17] He rejected Sadi Carnot's notion that no heat was lost. In his presentation, Clausius invoked Regnault's research on the properties of gas and steam in order to propose a mathematical conception of Sadi Carnot that was compatible with Joule's experiments.

Deeply impressed by the conclusions of Rankine and Clausius, Thomson nonetheless judged the latter's demonstrations to be inadequate. Following the logic of his analysis of mechanical effects, he managed, in an article of 1851, to combine the principles of mutual convertibility of heat and mechanical effects with the idea that losses of energy resulted. In this process of "dissipation," heat was not eliminated but wasted. Thomson thereby demonstrated the utility of Sadi Carnot's mathematical method, which alone made it possible to construct a complete theory of motor force that accorded with the calculations by Joule and Regnault. He proposed a new mathematical version of Sadi Carnot that explained both Joule's experiments and Clausius's hypotheses, while rejecting Rankine's vortex theory.

Thomson's synthesis cannot be separated from the evolution of his scientific thought or his close relationship with the universe of Glasgow engineers interested in the development of steam energy applied to naval construction. Thomson's initial education in mathematics and his adolescent admiration for the "splendor" of Fourier's work constituted one of the essential components of his ultimate achievement. An understanding of Thomson's achievement cannot ignore his participation in a milieu dominated by enthusiastic belief in the positive character of the mechanics industry and the conviction that there existed "an intimate connection of natural philosophy with practical arts." The unity of science and technology only illustrated the relationship "of mind to the material world."[18] This unity found another expression in the 1850s with Rankine's nomination to the chair of engineering at Glasgow and the publication of his book in 1859. This volume was intended to offer the reader "a credible new science which would serve as an infallible guide to the designing of more efficient and compact prime movers, especially for the marine use." Rankine added: "Thermodynamics would thus draw credibility from hot generating engines, and hot engines would conversely derive credibility from thermodynamics."[19] It was Rankine's work that consolidated the terminology of thermodynamics and confirmed the validity of a distinction between the first and second laws of thermodynamics introduced in 1857 with an article by John P. Nichol in the *Cyclopaedia of the Physical Sciences*.

Experimental Thermodynamics

Thermodynamics opened vast perspectives for operations of the steam engine and consequently for possibilities to improve its efficiency. In 1880 Robert Henry Thurston, an eminent engineer and president of the American Society of Mechanical Engineers, announced that Rankine's theory of thermodynamics was not comprehensible to professional engineers and therefore could not be useful for them: it remained too abstract and corresponded more to an ideal than to a real machine.[20] This somewhat polemical verdict translated the situation in the 1850s but not that of the 1880s and 1890s. It accounted for the fact that, until the end of the century, not all builders of machines had the same worries about improvement as did those who furnished sophisticated engines. The latter owed much to the science of thermodynamics, whereas most of the former were content to reproduce cheaper old models. The major factor remained, however, that the spread of thermodynamics through writings like Rankine's awakened a feeling that "something more than mechanical improvement was necessary to develop machines." Engineers realized that the steam engine was not solely a mechanical device but also "a machine in which heat is employed to do mechanical work." The principal objective of research should be to measure what was called the "missing quantity," that is, the gap between heat produced by a boiler and the energy delivered to the motor, then to find the reason for that difference, and finally to modify the machine accordingly.[21]

As noted, the question of a "missing quantity" remained unresolved. In his manual, to portray the four steps of a machine's cycle, Rankine had "replaced the real diagram with a simpler fictional diagram that resembled it."[22] His presentation thus retained a theoretical character. Yet it was not inconsequential, since it provided a conceptual framework within which experiments providing increasingly precise and convincing measurements could be achieved. But because patience was required to reach satisfactory conclusions, uncertainties were not immediately alleviated by the fledgling thermodynamics. Its theoretical character appeared to keep this science from successfully answering questions, though it helped to close the gap between theory and practice. This reconciliation was to be simultaneously accomplished by the Alsatian engineer

Gustave-Adolphe Hirn in France and by the American naval engineer Benjamin Franklin Isherwood.[23]

Hirn worked in the textile factory of Logelbach, founded in 1772, and eventually became its owner. In 1842 he was assigned to operate Woolf's high-pressure machine: "It was his industrial experience that motivated Hirn's research."[24] He wanted to conduct an "experimental theory" with the steam engine—to use the title of a book he re-edited several times—but his inspiration was very different from that of the Scottish school. The current of experimental theory had emerged in France in the 1840s at the École Centrale, where Léonce Thomas was teaching. Thomas had worked beside the industrialist Camille Laurens, with whom he had demonstrated the utility of steam casings—contested at the time—with experiments that made concrete analyses of changes that occurred in cylinders, such as condensation during the introduction of steam or evaporation during its expulsion.

Hirn established for the first time that condensation occurred during the expansion of steam, which had been only a supposition of Clausius and Rankine. To bolster his early experimental findings, he concentrated his efforts after 1862 on studying the effects of cylinder casings, whose importance had been underestimated. At the same time that Hirn was completing his research, between 1864 and 1868, the American engineer Benjamin Isherwood conducted a series of experiments at the Brooklyn Naval Yard. Begun in 1859, they sought to elucidate the effects of condensation in cylinders, high pressure, and the speed of motion in general. Like Hirn, Isherwood spoke of the phenomena of condensation and re-evaporation, which became the principal subject of investigation for "thermodynamic engineering" in the United States and Europe alike. The future possibility of achieving progress on the basis of both thermodynamics and experimental theory became considerable. According to Hunter, "as late as 1860 the best American engines consumed some twenty-five to thirty pounds of steam per horse-power per hour, whereas the theory of the mechanical equivalent indicated a consumption per horsepower per hour of only two pounds of steam." The framework and the principles that allowed existing procedures and practices to be evaluated and corrected and then to increase the production of heat in the boiler while reducing the loss due to friction in the machine were henceforth clearly defined. Engineers could better understand, and hence analyze and improve, new experimental procedures such as high pressure, compounding, and steam casings. Rankine in his texts thus proposed a precise analysis of liquefaction and re-evaporation and defined a series of concrete procedures able to offset the negative effect of valves. As Hunter has indicated in regard to the United States, "overall the impact of thermodynamics was less to change than to strengthen by correction and refinement the long-standing trends in this country toward higher steam pressure, higher piston and rotating speeds, and greater expansion."[25]

According to Thurston, the steam engine ceased to be an object depending solely on mechanical ingenuity and entered an era of steam engineering.[26]

The Conquest of Great Efficiency

Greater efficiency in steam engines was obtained through an increase of high pressure and the reduction of heat losses. These two factors determined all conceptions of engines. To adopt high pressure and use motor power optimally, it was necessary to invent devices that allowed the deployment of the laws of thermodynamics in a variety of circumstances. More generally, it was necessary to vary the valves, to master the increasingly complex and precise processes of distribution of steam in the cylinder, and to adapt parts and mechanisms to an ever more rapid movement of pistons. To meet these challenges, engineers had to gather know-how from their experience but also to collect theoretical and mathematical learning contained in countless treatises devoted to steam engines and general mechanics. Many engineers displayed brilliant imagination in responding to these needs. Others were deterred by the complexity of such new procedures and preferred to improve old methods, especially in the case of low-power machines for which the actual effects of innovation on productivity were not sufficiently important to compensate for the costs of investment and maintenance.

The Choice of Systems: Multiple Expansion

The major innovation between 1860 and 1900 was multiple expansion, double at first, then triple and quadruple. Around 1850 Arthur Woolf's engine was adopted in Great Britain by a limited number of industrial firms. Its development was conditioned on the acceptance of high pressure, though until then no theoretical argument had been advanced in its favor. High pressure was introduced in marine motors, locomotives, and textile machines, industries where an enhanced force for starting them was necessary. However, some textile regions did not adopt it. The millwrights of Lancashire continued to prefer a simple low-pressure wooden-beam pump or to operate two machines at once. John Elder was the first to employ a compound system for marine motors. After 1862 the use of compound motors in this sector spread rapidly, while their size and power constantly increased. For stationary machines this process was slower and less widespread. In the United States, the early adoption of high pressure allowed the use of valves thanks to the employment of variations controlled by a regulator. According to a census in 1900, four-fifths of engines constructed the year before had one cylinder, used slide-valves, had an average steam thrust of thirty horsepower, and functioned at rather low speeds that limited the effects of vibration. For such engines conserving fuel was not of great importance compared to the costs of maintenance. Single cylinder engines also continued to be generally used in Europe, and they even had a comeback on the two continents during the 1890s.

The new engines arriving in leading sectors such as metallurgy, electricity, and assembly line mass production were submitted to increasingly constraining needs for precision, mechanical complexity, regularity, and uniformity that typically required greater power. Three types of responses to these new industrial demands were mounted by increasingly professional engineers possessing ever more formalized and structural techno-scientific knowledge: the adoption of multiple expansion and high speed, the automatic regulation of movements, and better manufacturing quality in both materials and machines.

From the late 1860s the technology of compound engines became dominant in naval construction. It developed more slowly with stationary machines. In Europe the latter appeared at the London exposition of 1862, for the most part horizontal engines with weak power that increased in the 1870s. In the United States the American Society of Mechanical Engineers organized a symposium in 1880 specifically dedicated to this technology. One of the participants asserted without contradiction that "sooner or later compounding will not be confined to the narrow field of marine work, or to pumping, but will seek an outlet in the direction of manufactories."[27] It was particularly well adapted to the needs of electric power plants, which were somewhat comparable to naval motors. These machines could attain greater pressure and power. Horizontal engines were joined by vertical ones, of which the most powerful had four cylinders arranged in pairs. The strongest machine displayed at the Paris exposition of 1900 was a horizontal engine manufactured by the Escher Wyss firm. Thereafter much more imposing installations were constructed in the United States and Europe. Henry R. Worthington, one of the first American builders, stated in 1880 that "the compound engine is a way to get a high rate of expansion and to preserve a necessary uniformity of motion."[28] The English engineer Arthur Riggs, for his part, confirmed that it reduced the effects of friction and vibration.

In the 1890s these four-cylinder compound engines gradually ceded to triple or quadruple expansion engines. The first of these originated in 1871 with a naval engineer from Le Havre, Charles Normand, and in 1874 with the Englishman Alexander C. Kirk. After several tests starting in the 1860s, construction of stationary triple expansion engines began in earnest in the early 1890s. At first they were horizontal machines, but later vertically inverted engines appeared, reducing the floor space needed. Their functioning was exceptionally quiet and regular, perfectly suited for large textile mills. One 1,000-horsepower machine installed in 1905 at Ashton-under-Lyne continued to operate with the same regularity until 1966. Quadruple expansion was also tested, but its utilization remained limited principally to pumping stations in the United States.

Choice of Systems: High-Speed Engines

The first high-speed stationary engine was the result of cooperation between two engineers, Charles T. Porter and John F. Allen. Displayed at the London exposition of 1862, it was a double-action horizontal machine with four valves. According to Louis Hunter, this high-speed engine "occupied less space, consumed less steam, simplified problems of transmission, in many cases making possible a direct connection to machinery it served and reducing somewhat the costs per unit."[29] The tradeoff was that its functioning had to be close to perfection. Construction, precision in moving parts, balance of rotary blades, and regularity of motion required "zero error." This engine thus marked a major step in the creation of mechanical knowledge. Its system allowed considerable increase in the rapidity and continuity of the movement of valves. The engine could be operated like a locomotive. It produced a power several times superior to that of conventional engines of the same size while eliminating the heavy flywheels and most of the gears and cogwheels needed to regulate speeds. It enjoyed a rousing success at the Paris exposition of 1867, after which its uses multiplied. Compared with other machines of that era, the Porter-Allen engines created the same power while occupying a tenth of the space. They created enormous savings in the normal expenses of investment and maintenance. Moreover, the regularity of these engines was clearly superior to others.

Other models of high-speed engines were created in Europe, usually to drive the dynamos of electrical power plants. They were generally single-stroke machines such as the Brotherhood engine that appeared in 1871–1872, which comprised three cylinders arranged in a star. Because of its high consumption of steam, its use was ordinarily limited to installations in which a source of steam was already available. It was initiated by Charles Parsons, inventor of the steam turbine. For his part, George Westinghouse also invented a single-stroke engine that was suited to the electric lighting of ships. The machine most used in electrical power plants in the early 1890s was the Willans single-action vertical engine with triple expansion, which was also installed in textile factories.

The engines conceived by Sebastian Ziani de Ferranti for the London Electric Supply Corporation and then for the electrical power plant of Deptford, southeast of London, marked a new stage in the conquest of high speeds. At Deptford Ferranti at first installed compound vertical engines on opposite sides, functioning in tandem. He pursued this concept in 1895 in his own factory and succeeded in obtaining speeds distinctly superior to 200 revolutions per minute.

The outset of the 1890s was marked by a return to double-action engines. Those invented by George Edward Belliss were vertical machines with superimposed cylinders. The French construction firm Delaunay-Belleville had acquired a patent for a triple expansion engine, which it presented at the Paris exposition of 1900. It had four cylinders in two tandem groups with a speed of 250 revolutions per minute. These machines marked at the same time the zenith and the end of such high-speed engines, which were in reality without a future. Their actual market was small electrical power plants where economies were scarcely an issue and where they were soon replaced by electric or internal combustion engines. In the large centers they were outmatched by steam turbines. As Jacques Payen has concluded, "they never achieved enough savings to have a serious chance of survival." Yet their role was not negligible, since they opened "the way to the combustion engine, whose development would not have been so rapid if there had not been during the thirty previous years of the century solid and rapid engines that could be adapted to a new mode of operation."[30]

Variable Valves and Their Regulation

After the 1850s many small installations in cities continued to use simple low-power engines without their owners attaching real importance to the conservation of steam and fuel. Large and very large firms meanwhile appeared in great number. They needed to be equipped with much more powerful engines, consuming less fuel and assuring greater uniformity of motion, because the ultimate quality of products depended on "the regularity of their speed," to quote an American technological dictionary of the time.[31] Still more essential was the attention given to regular functioning of engines and the flow of steam during different phases of a piston's stroke, which varied between zero at the start or finish and a maximum in the middle of it.

In response to these needs, systems of variable automatic valves were invented in mid-century. The most successful of these was perfected by the American George Henry Corliss, who, after working as a clerk in a mechanical building firm in Providence, Rhode Island, became a partner in it. In 1847 he built a vertical engine largely inspired by those of steamboats. They were equipped with a variable valve device that he continued to improve, filing his first patents for it in 1849. Both his contemporaries and historians have contested his originality in this discovery, however, as new methods had emerged in Europe well before 1849. In 1823 Woolf had patented a system considerably improving the flow of steam, one of the major innovations assuring the success of the Cornwall engines. In 1836 the Frenchman Marie-Joseph-Denis Farcot patented a system of variable valves. In 1868–1873 his son Joseph invented the servomotor. Based on the principle of retroaction, it was designed to regulate the rudders of a ship and became a fundamental element of mechanics in the twentieth century. The work by Corliss thus entered into a general current of research on methods of control, automation, and functioning of steam engines. It was more a question of collective than of individual innovations. The most powerful electric generating machines at the 1900 Paris exposition were all driven by Corliss engines, constructed and modified in Farcot's factory, which won the grand prize.

In the United States the originality of Corliss was challenged by two of his fellow engineers, Frederick E. Sickels, who submitted a patent in 1842 for a variable drop-cutoff valve, and by Zachariah Allen, who in 1834 submitted one for a governor-controlled cutoff. The courts ruled in favor of Corliss on the grounds that he had been the first to combine the gravity valve with an automatic regulator. Hunter goes so far as to claim that in the history of steam engines, the mechanics of valves and their regulation invented by Corliss had an importance comparable to the inventions of Newcomen and James Watt. In fact, thanks to this machine, it was possible to radically modify the function of valves. The Corliss system allowed a much more efficient regulation of speed and power by varying the injection of steam during each cycle of the machine. Hence the simple valve was replaced by the variable valve.

The Manufacture of Machines

Despite their huge size, Corliss machines were conceived as part of a general attempt to improve the manufacturing of parts and resolve the problem of gaskets. The need for the utmost precision in the sizing and formatting of parts and their surfaces was the result of the growing complexity of distributors and valves. According to legend, this was an extension of the difficulties James Watt had encountered in obtaining a proper fit for cylinders and pistons. Thereafter the steam engine required nearly perfect precision of its components if one hoped "to control its dynamic force and minimize friction and vibrations that might seriously affect its practical use."[32] In 1775 John Wilkinson offered a response to this particular need with a machine capable of boring a cylinder of the desired size. In 1881, in the fifth edition of the *Dictionnaire des arts et manufactures* directed by Charles Laboulaye, the author of an article on tooling machines stated: "It was above all the necessity of creating parts of unusual size required by the construction of steam engines ... that led to an attempt to use purely mechanical procedures." According to this same source, these procedures began in Watt's workshops. The steam engine thereby became the fulcrum of the industrial revolution.[33]

This scenario turned out to be much more complicated. In the nineteenth century it soon became apparent that neither the foundry nor the forge could satisfy the demand for quality of the essential parts of the steam engine.

Products from the foundry were subject to contractions and frequent distortions while cooling, and blacksmiths could only approximate the form or size of a part. No one piece made by a worker could be exactly identical to another. To achieve a final shaping it was necessary to work at room temperature with hand tools. Well into the nineteenth century, machine manufacturers therefore called upon a disparate group of workers specialized in the metal trades. The most prized of them were the polisher and the fitter, but also useful were the finisher and the turner. Some parts required the services of a coppersmith, ironmonger, or bronze worker. These workers used very diverse tools and also some machines utilizing human or animal energy. The results depended entirely on the experience and know-how of the workers. Such methods were still largely employed in the 1830s, entirely so in the United States and in Scotland. Two decades later they were much reduced, but they did not completely disappear until the very end of the century. After the 1870s machines appeared that were characterized by the diversification and specialization of their functions, by their progressive automation even before the use of electric motors, and by the extension of their capacity and performance, whether due to the materials incorporated, the exactitude of shapes, or the precision of moving parts. The results Corliss and Porter-Allen had obtained in the United States with variable valves and high speeds could only be achieved through the quality in shaping and treating parts. In late nineteenth-century America, machine tools operated within a tolerance of a thousandth inch, thereby demonstrating steam engines' total dependence on the tooling industry of that era.

The End of an Era

Two conclusions clearly emerge from this analysis. On the one hand, the construction of technical knowledge, such as described by the example of the steam engine, cannot be reduced to a simple confrontation between theoretical science, which leads to "applied" science, and empirical technology, which aspires to an "engineering science." That would be an impermissible simplification of a much more complex reality. On the other hand, the construction of new knowledge was for the most part a product of innovations conceived by engineers and technicians who became entrepreneurs, or by intellectuals and engineers working in close collaboration with enterprises. The history of steam engines confirms a developmental model of technological learning that can be called entrepreneurial, since its dynamics depended on the initiative of entrepreneurs in competition to satisfy market demands.

In the system of knowledge elaborated around the steam engine, several trajectories unfolded and emerged without ever composing a total unity. By simplifying a bit, one may define three systems of learning created by these trajectories, which by combining could form for each of them an innovative dynamic. The first was thermodynamics. Its development derived from a confrontation between British natural philosophy and French theoretical mechanics, incorporating the results of experimental research founded by Lavoisier. The second system rested on a very early development, especially in Cornwall, of a statistical and experimental approach to the study of steam engines, the purpose of which was to define the procedures that were most economical, most certain, and best adapted to needs. This orientation flowed within a general current of forming engineering sciences such as applied mechanics and "railway science." The third system consisted of a slow but inexorable deepening of artisanal know-how and practices in the realm of mechanics and metallurgy. The principal source of this last type of innovation was a need to respond immediately and often urgently to the malfunction of machines.

Since the mid-nineteenth century, these three systems of knowledge have attempted above all to place the construction and function of machines into conformity with programs for which thermodynamics provided a framework. One of the main ambitions of the founders of thermodynamics was to furnish engineers with a program of coherent action, to which they came only belatedly. To satisfy the sometimes obscure recommendations of theorists, engine builders had to appropriate knowledge from an ancient tradition of workers and artisans who were independently capable of radically transforming and promoting new practices adapted to steam power. For this reason the concept of applied science is not apt in the case of steam engines, because it does not take into account the complexity of exchanges among the various forms of knowledge and the different actors in this drama.

The processes developed by each of these trajectories were not always coordinated. Uncertainties and hence controversies remained quite lively until the 1840s. They became somewhat attenuated by the 1850s, without entirely disappearing. At the end of the century a sort of coherence based on some apparently solid paradigms became established, and competition among the systems diminished. It was precisely at that moment, however, that the limitations of the steam engine driven by pistons were exposed in light of the emergence of steam turbines and the takeoff of electricity.

Chapter 7

The Chemical Industry

For the years between 1880 and 1960, it is possible to distinguish three types of technological culture that were evolving in different combinations. The eldest was the culture of workshops, inheritor of industrial practices in the eighteenth and nineteenth centuries. It survived two successive waves of industrialization and was still quite active in an enterprise like DuPont in the United States in the years from 1920 to 1950, the time of nylon and the atomic bomb.[1] At the other end of the spectrum was the scientific culture of intellectuals whose knowledge had served chemical firms since the beginning of the nineteenth century, supplemented in the last third of that century by fundamental or applied research in laboratories attached to such enterprises. A third culture combined an empirical approach with the engineering science taught in technical schools. Thus, along with the scientifically based laboratory research that established the foundation of a "new science," a specific know-how was created that followed the methodology espoused by Ernest Biver (brother of Hector Biver) in the aforementioned glassware factory of Saint-Gobain. Empirical learning thereby became formalized, and what was known in the Anglo-Saxon world as chemical engineering gradually became an autonomous discipline that was taught in the United States before World War I.

Four episodes in the history of chemistry illustrate the dynamics of exchange among these various types of knowledge:

1) the relationship between organic chemistry and the dye industry between 1860 and 1913;

2) the emergence of physical chemistry, in both the scientific sphere and industrial practice in the 1890s;
3) the simultaneous development of an industry employing colloidal products and polymeric materials, which touched off an intense scientific debate that opposed classical organic chemistry to innovative physical chemistry;
4) the renaissance of classical organic chemistry marked by the triumph of macromolecular chemistry that sought to integrate physical chemistry into its theoretical structure.

The plastics industry was born through these controversies and conciliation. The different episodes in this process demonstrated the importance of controversy in the evolution of sciences and the appearance of new technologies.

Organic Chemistry and the Dye Industry before 1900

In the realm of organic chemistry, the discourse among different types of learning followed a rather different trajectory than that of steam engines and thermodynamics. The epistemological barrier between various approaches was more quickly and simply overcome because a kind of natural affinity existed between scientific laboratory research and industrial practices in several sectors of activity, such as the production of gas or dyes. In the course of the nineteenth century, these were the first applications of the classic organic chemistry of crystalloids. Spurred by research into artificial products that could be substituted for natural ones, this process was extended to other sectors of activity such as polymers, which spawned the plastics industry.

The Birth of Organic Chemistry

The first step in this history began in the 1860s with the affirmation of organic chemistry as an autonomous discipline in the science of the nineteenth century. This was the result of developments, both theoretical and methodological, that had begun at the outset of the century, originating in the chemistry of Lavoisier and John Dalton. Lavoisier's nomenclature furnished chemists with a common language and a coherent definition of chemical reactions by using weight as an instrument of quantification. Dalton had asserted that each substance was formed by different types of atoms and showed that the relative weight of atoms could be measured. Measurement of atomic weights and theory about organic components would dominate research for generations to come. Even though most chemists declared opposition to atomic theory, they pursued their research within that context. To borrow an expression of Jöns Jacob Berzelius, chemical research, which powerfully accelerated in Germany in the 1820s, was oriented toward "the composition of substances and of their relationship to one another."[2] Such an approach eventually created a separation between chemistry and physics. In an essay published in 1818, Berzelius offered corrections to Dalton's atomic weights. In order to determine atomic weights, chemists "mobilized all available resources, knowledge as well as know-how, from the study of gas to crystallography, including the theory of heat and various techniques of measurement."[3]

Berzelius also introduced the terms "isometrics" and "polymetrics," thereby opening a debate on the different possible arrangement of compounds. Polymers such as ethylene were defined as compounds combined in identical proportion but constructed by a different number of atoms. It was the controversy about varying possible interpretations of the process of combination and substitution that laid the groundwork for organic chemistry between the 1820s and the 1860s. The French—particularly Jean-Baptiste Dumas, Charles Gerhardt, and Auguste Laurent—made essential contributions to these debates. In Germany, Justus von Liebig, a student of Gay-Lussac, was named professor at the University of Giessen, "a small university in an unimportant state," where he created a laboratory to determine the constituent elements of a multitude of products. He played a major role in forming a structural theory of organic chemistry, supplying a more precise definition of the concepts of "radicals" and "isometrics."[4] Following the successful synthesis of urea by Friedrich Wöhler in 1828, the most prominent objective of chemistry became the production of artificial organic components. This synthesis was the first of a long series realized after the 1850s.

Wöhler's discovery "openly posed the question of atomic arrangement," and it was no longer possible "to be content with characterizing a compound by the nature and the proportion of its constituents."[5] Several researchers in the 1850s proposed the concept of "valence," or the value of combination, which attempted to explain why different elements combined in certain ways rather than others by simply attributing to atoms a unique capacity for combination. August Kekulé von Stradonitz, a student of Liebig and Charles Gerhardt, preferred the word "atomicity" to describe this process. The concept opened new paths of investigation, because it was "an instrument of prediction as well as a program of research." The resulting chemical formula could specify the number of valences exchanged among different constituent atoms and "thus indicated the ways in which a synthesis could be achieved."[6] A new threshold was crossed in 1858 with Kekulé's discovery of the tetra-valence of carbons, which contained the entire future of synthetic chemistry. Working with methane gas, he demonstrated that the atoms of carbon could be combined and could exchange a valence. This finding explained the chemical peculiarity of carbons and confirmed a law of Count Amedeo Avogadro, announced in

1811, that made a distinction between atoms and molecules. This law had been rejected by a majority of chemists before the 1850s, but in 1860, at a congress in Karlsruhe, a consensus was reached on the difference between the two types of particle. The concept of isometrics advanced by Berzelius gained full meaning, and it became increasingly clear that the properties of substances depended on the arrangement of atoms rather than their number. Kekulé's work on the distinction between aromatic and aliphatic substances demonstrated the wealth of possibilities opened for research based on analysis of molecular theory. A new effort was mounted to discover and artificially synthesize organic substances. The number of organic compounds grew from 3,000 in 1860 to 20,000 in 1883, and to 140,000 in 1910. Starting in the 1870s, this explosion was particularly spectacular in Germany, where most of the newly created laboratories were devoted to synthesizing materials that could be crystallized or distilled. The instruments of scientific inquiry made available via organic chemistry derived from the theories of Kekulé and his disciples allowed a definition for the specific properties of each synthesis and offered virtually limitless possibilities for producing new substances that could have a variety of industrial uses.

Colorists and Chemists before 1870

Research into the process of coloring had reached an impasse in the second half of the eighteenth century. Only in the control of quality had science had begun to find improved procedures. In the course of the nineteenth century, however, a cluster of information formed as close connections were established between colorists with increasingly refined formalized learning and ever more numerous chemists with academic training acquired at universities where chemistry was taught. Robert Fox and Agusti Nieto-Galan have analyzed this process as follows: "The complexity of the procedures, the scientific discussions that were involved, and the assimilation of a huge range of exotic raw materials steadily transformed an atomized world of practitioners, with traditions rooted in guild secrecy, into a network of experts between whom knowledge and skills in the preparation and use of natural dyestuffs circulated rather freely."[7] Within this network of experts, personal and professional ties became increasingly tight. Visits and meetings, academic debates, and publications maintained these contacts in a turbulent world subject to the pressure of an expanding textile market and strong internal competition. Business enterprises were the principal instrument of these confrontations, in which the main weapon by the 1860s was the patent. German chemists and chemical firms deployed it better than others, outdistancing extraordinarily incompetent French and English courts.

Colorists were at the center of the network. In the 1850s and 1860s they were "sought out, highly paid, and respected."[8] They were most numerous in France, especially in Mulhouse, Rouen, and Lyon. In Germany and Great Britain they were to be found in great centers of the textile industry such as Elberfeld, Krefeld, and Manchester. Many of them had been educated in special schools like the École de Chimie at Mulhouse, created in 1822 by local entrepreneurs, or the schools of Elberfeld and Krefeld. Heinrich Caro, future director of the research laboratories of BASF, was trained as a colorist at the Gewerbeschule of Berlin. The establishment of such instruction illustrated the transformation of tacit artisanal learning into formalized expert knowledge. Some colorists like Camille Koechlin received academic training in chemistry, and most of them had intermittent or sustained contacts with universities throughout their career. Caro began as a cloth printer. He acquired such a sterling reputation as a colorist that his advice was sought everywhere, particularly once he had also mastered the techniques of manufacturing and marketing natural dyes. Possessing these diverse skills acquired on the job, he took up relations with August Wilhelm von Hofmann and with chemists in general. In fact, German universities like that of Giessen, which had created relatively significant programs for education in chemistry, did not have a sufficient number of academic posts to employ all the chemists they trained. Industry offered these students such attractive opportunities that young chemists also became colorists.

In France, university chemists like Michel Eugène Chevreul, Jean-François Persoz, and Paul Schutzenberger maintained close relations with the dyestuffs industry. Hofmann arrived in England from Germany to become head of the Royal College of Chemistry from 1845 to 1860. He attracted several English and German colorists, among them Caro, and articulated the relationship between chemical science and industrial practice, thereby assuring a greater autonomy to the former than it enjoyed in textile centers. Once returned to Berlin, he exercised considerable influence in Germany. Hence the two spheres of chemists and colorists became "intimately connected" everywhere in Europe.[9]

The Transition from Natural to Artificial Dyes

The trail that by the century's end extended to a mass market of synthetic dyes was the result of integrating two bodies of knowledge and research. One was the learning acquired and research conducted in the dyestuffs industry by colorists who used natural products, animal or vegetable, as primary materials. The other was knowledge gathered from chemists' research on products derived specifically from coal tars, supplemented by several other organic products. The first discoveries in tar chemistry were mainly due to the students of Justus von Liebig, who conducted many systematic analyses of this miracle product. Among their discoveries were naphthalene in 1820, anthracite in 1832, and carbolic acid in 1834. Colorists soon became interested in these discoveries because tars contained nitrogen, whose properties were similar to those of alkaloids derived from plants. Thereupon research took three directions: to find new dyes responding to fashion trends, to improve the stability of colors, and to reduce costs of production in order to promote the democratization of these new dyes.

Whatever the trend, it quickly became apparent that the products of organic synthesis offered much greater possibilities than those derived from natural substances. In this war of dyes, different strategies competed: as in the case of French and English chemists, to create new products without taking into account the nature of the chemical substance, natural or artificial; to explore the constitution of dyes extracted from tars, especially aniline, following the procedure used by Hofmann in London and then Berlin, and subsequently by his German disciples; or to gather the research (and researchers) from elsewhere and to improve their results, as the Swiss did. Whereas French procedures remained dependent on empirical findings, the approach of Hofmann and his German successors was more "systematic," as Adolf von Baeyer later put it. The latter unfolded in three phases: defining the composition of natural dyes long utilized by the industry; determining the constitution of artificial dyes accidentally obtained in previously conducted experiments; and, once those chemical components had been established, achieving an industrial synthesis of the dyes tested.[10]

Three steps in the history of dyes can be distinguished for the period from about 1850 to 1880. The first consisted of important advances in both natural and artificial dyes, realized by the appearance of large-scale products and the first dyes extracted from tars. This discovery opened the way to a second stage, mostly Anglo-French, characterized by the takeoff of aniline dyes and a weakening of the French position compared to British successes, largely due to Hofmann and his network. Then, in a third phase, with Hofmann's return to Germany and the migration of several French colorists to Switzerland and Germany, German business enterprises and universities took the lead.

In his "Leçons de chimie appliquée à la teinture," published in 1830, Michel Eugène Chevreul defined the principles of dyestuffs and described the process of their corrosion. Parting from this basis, French and English colorists made several remarkable advances in the 1840s and early 1850s, in natural and artificial dyes alike. At Lyon in 1845, the colorist Jean-Aimé Marnas developed a yellow dye from picric acid, which he produced in 1849 with resin obtained from an Australian plant using a procedure suggested by the chemist Auguste Laurent, who was in contact with the Manchester chemist Frederick Grace-Calvert. In 1847 the latter, who had received academic and industrial training in France, succeeded in producing from tars a phenol (carbolic acid) of excellent quality, from which he extracted picric acid. Grace-Calvert supplied this product to Marnas, and a factory producing picric acid was opened near Lyon in 1855, directed by a former chemistry professor, François-Emmanuel Verguin. This dye gave a magnificent color, but it was unstable and could not meet the competition of aniline after 1863.

Marnas also developed a resistant dyestuff called French purple, produced from the common European lichen *orseille* or "dyer's moss" by combining the reaction of air with ammonia. Other dyes of this type were created, derived from different kinds of moss and furnishing a varied palette of red and blue colors. One of the most successful, because of its stability, was perfected by John Stenhouse, a lecturer in pharmacology at St. Bartholomew's hospital in London. Marnas adopted this method in 1857 and devised a procedure for dying not only silk and linen but also cotton. For his part, Paul Schutzenberger, a student of Jean-François Persoz (then professor of chemistry at the school of applied chemistry at Mulhouse and later professor at the Collège de France), launched research in the mid-1850s on dyestuffs produced from madder-root. He achieved this goal in 1869, a date simultaneous with the appearance of alizarin dye, which was to compete with and soon replace these other products.

Murex dyes, produced from uric acid, had been studied since 1838 by Liebig and Wöhler. In 1853, Dr. Frédéric Sacc and Jules-Albert Schlumberger were able to extract uric acid from the excrement of wild boars and to obtain from it a brilliant color. The Parisian enterprise Depouilly Frères achieved the same result from Peruvian guano and secured a patent for dying silk. In 1856 Charles Lauth of Strasbourg, also a student of Persoz, patented a corrosive allowing the application of uric acid to cottons. In 1856 the firm of Nicolas Koechlin et Frères produced and marketed murex dye. This production was expanded thereafter in large cloth-printing enterprises—French, English, Bohemian, and Saxon—reaching its peak in 1859. Its decline was rapid, however, because the color faded through contact with heating gas, whereas fuchsine (aniline magenta dye), soon to appear with a similar color, was much more stable. According to Ernst Homburg, murex dye was "the first dyestuff used on a large scale." Moreover, this product was "radically different from traditional natural dyes."[11] It opened the path to aniline dyes, since the manufacturing procedure used was no longer by extraction but synthetic. Thus, the science of natural dyes pioneered the way to artificial dyes.

The second stage in the history of dyes sprang from the commercialization of the first aniline dye in 1856. This discovery ensued from Hofmann's work in the 1840s at the Royal College of Chemistry in London: research on tars and their nitrous and amine components. In 1843 Hofmann uncovered the identity of an alkaline that could be extracted from tars and an oil obtained through the distillation of indigo, namely aniline, composed of two hydrogen atoms and one nitrogen atom. He wrote in 1849 to one of his patrons, Walter Crum, "that none of these components have as yet found their way into any of the appliances of life. We have not been able to use them for dyeing calicot nor for curing disease."[12] To overcome this obstacle it was necessary to take a detour into the analysis of natural dyes, breaking down complex molecules into smaller elements either by simple distillation or by procedures of decomposition. The discovery was achieved by one of Hofmann's assistants at the Royal College, William Henry Perkin. Recruited by Hofmann's laboratory at the age of fifteen, he was assigned to test new analytic methods in France and Russia in order to study the nitration of carbohydrates extracted from tars. By doing so, he hoped to obtain "the artificial formation of natural organic compounds" by creating a synthesis of natural quinine through treating a tar-based amine with the aid of an oxidizing agent. He produced an auburn color and then a solution of intense purple. Thereafter, Perkin wrote, "by experimenting with the coloring matter thus obtained I found it to be a very stable compound dyeing silk a beautiful purple which resisted the light for a long time."[13] To be sure, he still needed to convince the initially skeptical textile industrialists of his product's quality, but his success was immediate due to strong demand for the color purple, influenced by a fashion to which Queen Victoria herself gave a decisive impulse.

Perkin's success stimulated the cooperation of chemists and colorists, who multiplied their experiments of combining aniline with other available substances, engendering a real boom of aniline compounds between 1859 and 1865. François-Emmanuel Verguin in Lyon was among the first to obtain results, which were "half theoretical and half by chance."[14] He discovered a red dye that he christened fuchsine and sold his method to the firm of Renard Frères, which commenced production in November 1859. The process was improved by Louis Durand, who used mercury as a reagent. Several dyes of the same type as fuchsine and its blue or violet derivatives enjoyed considerable commercial success.

Hofmann's research owed much to his relationship with industrialists like Edward Nicholson, who furnished him with precise information about the experiments he was conducting in France with Charles Girard and Georges de Laire. These enterprises were also able to supply him with pure aniline. He directed research toward analysis of the chemical composition of aniline dyes, first red and then blue. He showed that red aniline, fuchsine, or magenta formed as the product of a mixture of aniline and toluidine, and in 1862 that blue aniline was the product of the phenylated red aniline. Hofmann's work was a major step in the transition from innovation based on a combination of trade practices and academic knowledge to innovation anchored primarily in science. There is no question that research associated with the chemistry of dyes exercised a direct influence on the evolution of chemical theory. It was in 1865 that Kekulé presented his molecular theory by applying the byproducts of coal tar to facts drawn from chemistry. Thus he defined the structure of benzene, thereby providing a basis of the structural theory of molecules that would dominate the world of chemistry for another century.

Hofmann's research did not focus solely on the preoccupations of science and industry. He was also much engaged in litigation, a battle of experts in France and Germany that strongly contributed to a deepening of scientific research at the end of the 1850s. "Certain chemists," as Anthony S. Travis has commented, "attempted to promote their ideas as much with legal arguments as with scientific facts."[15] In France, between 1858 and 1863, a trial ensued from a charge of fraudulent imitation brought by Renard Frères against other French manufacturers of red aniline, like Gerber-Keller. The court decided that Renard Frères was the only firm allowed to market the reactions of red aniline in France, declaring that this invention under a chemical patent involved both the actual product and all the procedures employed to create it, and holding that the new methods developed since fuchsine were identical. This decision showed the immense chasm separating the world of science and technology from the universe of legislatures and courts. Several French enterprises consequently relocated to Germany and especially Switzerland: Gerber-Keller, for instance, moved from Mulhouse to Basel. However, the firm of Renard Frères was unable to exploit its monopoly and, badly managed, went bankrupt in 1868.

Hofmann's career revealed a dual dynamic in the construction of technological learning in a milieu where the borders between technology and science were unclear. Empirical and purely technological findings led to questions to which science responded, but only with long delays and incompletely. Questions arose when technological facts were met by scientific uncertainties, and to adequately answer them, intellectuals had to rely on technological know-how provided by industry. Inversely, the perspectives opened by scientific discoveries could only be exploited by a refinement of theory or by provisional solutions.

Another path of research was investigated in England by Heinrich Caro, employed in 1859 by the Manchester firm of Roberts Dale & Co., which produced the mauve dye of William Henry Perkin. Caro extended research begun in 1858 by two chemists, Johann Peter Griess, a student of Hofmann at the Royal College, and Carl Martius. These investigations, without the use of corrosives, produced very stable dyes obtained by the effect of nitric acid on aniline. Caro industrialized this process at Manchester by "making azoic salts react with aromatic amines." In 1864 he created two azoic dyes, "which he named after their birthplace: Manchester yellow and Manchester brown."[16] Caro also invented the first dyes from naphtalene and carbolic acid. His first treatise on theoretical chemistry, written in collaboration with James Alfred Wanklyn, defined "the relationship between two important dyes extracted from tar, rosaniline (fuchsine or magenta) and rosolic acid, from which other colors were derived.[17] Caro's example did not remain isolated after other Germans such as Ivan Livinstein settled in Manchester.

Representative of a distant but symbolic third step in this history were the virtually simultaneous initiatives of three German chemical enterprises—Bayer (Friedrich Bayer & Co.), BASF (Badische Anilin- und Soda-Fabrik), and Hoechst—at the beginning of the 1860s. Thirty years later they would be leaders in the world market for dyes, since English pre-eminence proved to be of short duration. By the 1870s, the German industry of dyes and organic products in general loomed as the most dynamic in the world and rapidly seized the lead. Until the 1880s, this evolution was not due to the importance of German industrial research but rather to the effectiveness of academic research that was increasingly oriented toward dyes and other useful products. Most of the German chemists living in England, such as Hofmann and Caro, returned to their native country. The symbolic harbinger of the future triumph of German chemistry was the synthesis of alizarin, until then produced from madder, by two chemists, Carl Graebe and Carl Theodor Liebermann, during research at the chemical laboratory of Adolf von Baeyer at the Berliner Gewerbe Institut, a BASF-affiliated enterprise founded in 1861. In 1868 it recruited Caro, who had returned from England after seven years at Manchester.

From an institutional standpoint, this discovery symbolized the inception of a holy alliance between university and industry. As for the construction of knowledge, it was related to the contemporary invention of the Solvay process for manufacturing soda, which resulted from a long, persistent effort of research and development. This discovery was not "on the fly," as in the case of Perkin's mauve dyes, but was "anticipated, programmed, and the fruit of extended research."[18] It confirmed a procedure that was not entirely new in principle but became the foundation of planned and increasingly massive research, in terms of the quantity of men and materials as well as the number of operations. The synthesis of alizarin followed the same logic as those derived from aniline, since it was likewise a product of the distillation of tars from anthracene. Actually it was different, insofar as the necessary operations were complex, requiring additional research by Caro after 1869. Thanks to the use of sulfuric acid, he attained a commercial product in 1872 that met with immediate and total success, since synthetic alizarin provided an excellent substitute for madder-root, whose cultivation ceased after a few years. The subsequent emergence of BASF was founded on the success of this process. The profits realized from it provided the means to launch an active research endeavor through the creation of huge laboratories.

Given this scenario, it is possible to define an initial mode in the construction of knowledge that may be evaluated as both entrepreneurial and interactive. It rested on a dual complementarity: that which united the academic world of chemists with enterprises producing dyes, and that which associated professional colorists with university chemists. These business firms tended to be small, concentrated in large textile centers like Lyon, Mulhouse, and Manchester, and integrated in a market undergoing stiff competition. Information circulated rapidly in this universe, and technological and scientific experiments were at the heart of production. Hence one may speak of a system of collective innovation unfolding within an arena of global learning.

Birth of Vertical Integration: The History of BASF

In sum, the founders of the new chemical industry that emerged in the 1860s and 1870s, such as Hofmann and Caro, were excellent connoisseurs of both organic chemistry and the secret methods of vegetable dyes. They were capable of melding their academic knowledge with their industrial and commercial know-how. This was the basis, between 1880 and 1900, of the power of the German dyestuffs industry, which permitted it to establish its domination, with the Swiss industry alone able to mount significant competition. In 1913 Germany produced 85 percent of the world's dyes, a success explained in large measure by the massive effort of industrial research and development. That would not have been possible, however, if the German system of education had not supplied the scores of engineers and researchers necessary for such development. Beginning in the 1880s, the large German chemical enterprises moved from a personalized system, in which research depended mainly on one or another individual belonging to a firm or not, to "a system of institutionalized research" characterized by the creation of industrial laboratories, the appearance of a research infrastructure, and the establishment of positions for directors of research.[19] These laboratories did not in fact eliminate the need for individuals outside enterprises, whether engineering advisers or university intellectuals. The two systems, one personal and the other institutional, were thus superimposed.

The model identified here as vertical integration is well illustrated by the three large German chemical firms, BASF, Bayer, and Hoechst, which joined in 1925 to form the cartel of I.G. Farben. Bayer's origin was an enterprise founded in 1861 by Friedrich Bayer and Johann Friedrich Weskott, both involved in the dyes industry. It manufactured aniline on a small scale before launching production of alizarin in 1872. After Bayer's death it became a joint-stock company. In 1884 its director, Friedrich Carl Duisberg, founded a laboratory that was enlarged in 1891 and fascinated generations of researchers. Hoechst began with a chemist, Eugen Lucius, who cooperated with businessmen to found an enterprise in 1862. In the 1870s Hoechst became one of the leaders in alizarin production. Bayer and Hoechst, on the basis of their knowledge of dyes, concentrated in the 1880s on the research and manufacture of pharmaceutical products without totally abandoning dyes and other synthetics.

Caro pursued his career at BASF as director of its central laboratory until 1889. In 1874 the firm had merged with two other dye companies, which added an important marketing capacity. Alizarin-based dyes, which flourished in the 1870s thanks to intensive research efforts, represented half of BASF's business at that time. Caro's contract with the firm stipulated that his duty was to pursue his research in pure science and "above all to contribute thereby to the prosperity of society."[20] His laboratory employed eight chemists in 1883. Their research strategy was to analyze the dyes introduced by competitors. Between 1874 and 1883 Caro had a hand in 40 percent of the new dyes introduced by BASF. He worked in close contact with Adolf von Baeyer, professor at the University of Munich since 1875, and he established relations with nearly all of the university chemists of the time.

The cooperation between Baeyer and Caro notably concerned phthalic dyes, nitric compounds, and the structure of alizarin. It also included indigo, the most lucrative of all natural dyes, an investigation Baeyer had already undertaken in 1865. In 1869 he had foreseen a synthetic method for creating indigo, but he abandoned this research until Caro called his attention to it in 1877. The problem then posed was more theoretical than industrial in nature, which was why Baeyer was interested. He discovered the first synthetic indigo in 1878 and thereby defined the possible approaches of industrial research. Rather than obtaining a patent as Caro wished, he published his results, thinking his procedure would attract little notice because of the low price of indigo. In 1880, pushed by Caro, Baeyer derived a synthesis from cinnamic acid and tar that was much more easily applicable to industry, to be called methyl benzene. BASF purchased the patent, and Baeyer pledged his energetic support in developing the process. Although BASF joined with Hoechst to create a mutual research cartel in June 1880, the industrial process of synthetic indigo was not commercialized until 1897. The program was immediately launched in 1880, but collaboration between Baeyer and Caro proved difficult. Baeyer wanted first of all to establish a correct formula for the structure of indigo before attempting to implement production on a grand scale. Caro, thanks to his experience with synthetic dyes, was able to improve production of cinnamic acid. But in November 1880 the attempt to transfer laboratory procedures to mass production failed. Baeyer concluded that industrial efforts had been precipitous and that a deeper, more systematic preliminary scientific investigation was needed. For his part, Caro introduced a new method of dying cloth that was the first industrial application of artificial indigo. Yet this product had limited commercial use, and the program languished in 1883, along with the project of substituting synthetic for natural indigo.

Important changes in the management of BASF occurred in 1884. The new directors reoriented research efforts toward the discovery of "practical procedures," to cite the expression of one of them, Heinrich von Brunck, in explicit reference to the indigo program. Thus, a new vision of research emerged at BASF. Much of the activity of the central laboratory was absorbed into the administration of patents and put under Caro's supervision. In the final years of the decade, half of the central laboratory was working on patents. Hence research depended at least as much on the analysis of patents as on knowledge of scientific literature. Caro left the directorship of the laboratory in 1889 but maintained close relations with the firm as a consultant and member of the executive council. His successor, August Bernthsen, undertook not only to find new procedures and products to realize profits from scientific advances but especially to improve existing procedures of production and to support research programs in the two laboratories created in the 1870s, those of aniline and of alizarin.

From 1890 to 1914 the central laboratory of BASF was mainly occupied with responding to the industrial uses of dyes. The research was oriented to both the recognized demand for these products and knowledge of the molecular structure of chemical substances. It progressively evolved toward a study of physical properties as well as chemical structure, thereby resembling physical chemistry, a fashionable discipline in the 1890s. In 1899 BASF perfected a highly successful lacquer dye. Synthetic indigo was commercialized in 1897, thanks to research conducted in the departments of aniline and inorganic chemistry. Shortly thereafter, researchers at Hoechst developed an industrial synthetic indigo by utilizing a different, far less expensive procedure than that of BASF. Soon a compromise patent between them was signed.

After 1900 research at BASF was decentralized. The firm's activities became more diversified as a result of innovations like catalytic hydrogenation and synthetic tannins developed in the company's laboratory. The department of indigo conducted work on nitric acid and synthetic rubber. Liquidized chlorine, sulfuric acid, and particularly the synthesis of ammonia under high pressure—the most spectacular success—were elaborated by the department of inorganic chemistry, which became the "focal center" of research at BASF.[21]

All in all, the strategy of cooperation between university chemists and BASF, put into place in the 1870s, proved fruitful. At the end of that decade, the firm called on Adolf von Baeyer, the uncontested leader of the discipline, to be its scientific adviser and appointed him to direct a program of research. He had to return to more concrete realities in the 1880s by following the progress being realized by BASF's competitors, analyzing their patents, and developing more systematic scientific programs. Thus, he was primarily concerned with transferring laboratory procedures to industrial production. The failure of the initial program of synthetic indigo, an academic effort carried out within the central laboratory, redirected his research toward evaluation of existing market conditions. By the 1890s, research on artificial dyes had become "independent of academic chemistry as a day to day project at the sites of production."[22]

Henceforth institutionally independent, this research gained its own laboratory, defined its own objectives, promoted commercially successful products, and controlled its own publications and patents.

The history of synthetic indigo at BASF demonstrates the ambiguity of a conception of basically scientific industry. Although the initial programming had been inspired by recent advances in chemical theory on the structure of molecules, its ultimate form was achieved in laboratories of industrial research that combined the new discoveries of organic chemistry with the experience of industrial workshops and laboratories. Therewith entrepreneurial and interactive methods continued to function, but they were gradually transformed by the appearance of large enterprises whose research strategies took into account the perspectives offered by both the state of knowledge and market conditions. This policy was realized as the result of three main components. At the heart of the system were the laboratories, attached to different departments and consequently in direct contact with the production and commercialization of goods. They worked in close collaboration with a central laboratory devoted to more speculative and fundamental research. Finally, close relations were established with the university and the scientific community within a larger disciplinary framework.

This research model of so-called vertical integration developed while reinforcing organic chemistry as an autonomous academic discipline. A wide consensus was established, armed with the analysis of molecular structure. This allowed a definition of coherent research programs and constantly opened new territory upon the discovery of new substances and dyes. Universities were able to supply industry with all the young chemists needed to drive research programs, dye production, and market outlets. In the end, the history of the dye industry shows how the initial choices of research strategy determined ultimate trajectories. The German scientific community was very early oriented toward the priority of synthetic dyes, whereas the Anglo-French community long remained attached to natural dyes that were thought to result in better quality. Furthermore, Anglo-French scientists adhered to an organization of research that was more individual, less collective, and less massive.

Physical Chemistry in the Second Industrial Revolution

After the 1880s a wave of innovation similar to that of the years 1760 to 1870 spread across the Western world. It laid the foundation of the twentieth century by creating the technologies and sectors that would dominate its economy and society. The most spectacular of these were doubtless electrification, both industrial and domestic; the internal combustion engine; the wireless telegraph; and the "new chemistry," both organic and physical. This process may be designated as the second industrial revolution because of the radical changes it brought, compared to the technology of the first industrial revolution. Strikingly, this second period was contemporary with the emergence of new academic disciplines, which by the 1890s affirmed their autonomy by proposing new approaches for the study of both technological and natural phenomena. One of these was physical chemistry. Its development had considerable industrial repercussions marked by the discovery of synthetic ammonia and the transformation of the polymer industry into a basic scientific undertaking.

The Emergence of Physical Chemistry

In the 1890s research at BASF took a new course, keeping pace with the evolution of German academic chemistry. This field was then undergoing a major convulsion because of the takeoff of physical chemistry, whose advocates laid claim to an autonomous discipline following its development in the 1860s. It was to extend a Newtonian current of thought articulated particularly in the eighteenth century by Claude Louis Berthollet. During the resurgence of the 1860s the authors concerned aimed to explain the dynamics of chemical reaction and phenomena such as the effects of temperature and of high pressure, distillation, and catalysis. At first a number of them, like Marcellin Berthelot, created a new discipline, thermodynamics, describing the interplay of chemical reactions and heat. Yet at the time no durable relationship was established between chemists and physicists.

The idea that finally had the most success was thermodynamics. It was clearly enunciated in the 1870s in the work of the German August Friedrich Horstmann, who in 1874 published a theory of dissociation founded on the second principle of thermodynamics, and the American Josiah Willard Gibbs, who developed "a nearly complete thermodynamic system."[23] In France in 1883, 23-year-old Pierre Duhem submitted a thesis on thermodynamics inspired by Gibbs, only to be rejected because it was forbidden to contest the validity of Berthelot's notions. In Germany, to the contrary, three young intellectuals laid the foundation of a true discipline: the Dutchman Jacobus Henricus Van't Hoff, the Swede Svante Arrhenius, and the German Wilhelm Ostwald. They sought to utilize the laws of thermodynamics and kinetics to prove the dynamics of matter. This physical chemistry claimed to break with classical organic chemistry such as that practiced and taught by Baeyer. It is noteworthy that these three intellectuals began their careers outside of the university circles that dominated the German chemistry of that era. Their somewhat atypical education had permitted them to gain a better understanding of physics than was usual among the majority of German chemists. Van't Hoff became known for a study of the atomic structure of carbons. The first work of Arrhenius analyzed the nature of solutions and advanced a theory of electrolytic solutions inspired by Rudolf Clausius, dating from 1857. Ostwald attempted to clarify the concept of affinity by studying the action of masses. In 1887 he was named professor at the Technische Hochschule of Leipzig, where he attracted a great number of young German and anglophone chemists. At the same time, he and Van't Hoff created a scientific journal devoted entirely to physical chemistry.

Thus was formed a system of knowledge founded on a coherent body of isolated scientific facts and on a combination of kinetic and thermodynamic approaches to chemical reactions. Chemical equilibrium was defined in accordance with the second principle of thermodynamics. Van't Hoff's contribution was essential to this point of view, but he was part of an entire movement marked by the works of Duhem, Gibbs, and several others. In the late 1880s Walther Hermann Nernst, an assistant of Ostwald in Leipzig, proposed a system of equations to quantify the effects of electromotive force. It thereby became possible to predict, with a certain degree of confidence, chemical and physical properties and the level of dissociation of a great number of electrolytes. Hence it became clear that "the apparently disparate properties of matter were in reality closely related" and therefore quantifiable.[24] This chemistry made greater use of mathematics, subordinated laboratory experiments to theory, and erased the border between organic and inorganic chemistry.

For its expansion, physical chemistry possessed, both in research laboratories and industrial firms, increasingly effective instruments of experimentation and production. Between 1880 and 1914 the very nature of chemical observations was entirely transformed. The spectral analysis invented in the 1860s by Robert Wilhelm Bunsen made great progress. Richard Adolf Zsigmondy perfected the ultramicroscope in 1903. In 1895 Carl von Linde invented an apparatus to liquefy air. Electricity plants constructed around the turn of the century attained power unimaginable in the 1880s. The development of electrolysis provided important quantities of hydrogen. Finally, the discovery of x-rays created analytic instruments of a precision inconceivable until then and established a new science. It became possible to research increasingly low or high temperatures and pressures. Henceforth the limits of experimentation were more and more expanded, even as questions requiring quantified answers were spreading.

The relationship between scientific knowledge and industrial practice also changed radically. Scientific theory furnished engineers with increasingly precise instruments of analysis and prediction for constructing installations and directing production. They acquired equipment that was totally regenerated by the use of electricity, higher-quality materials, and the emergence of a new technological discipline, chemical engineering, which moved to classify and rationalize the processes of production while taking into account, at least partially, the evolution of chemical science itself.

Thereby, entirely new industries were born or grafted onto traditional practices. This development included electrometallurgy and electrochemistry as well as physical metallurgy and the industry of polymers. It is not possible here to enumerate all of these transformations, but two remarkable accomplishments of the chemical industry—the synthesis of ammonia by catalytic procedures and the first creation of synthetic matter—illustrate the compatibility established between the experimental dimension of physical chemistry and that of colloidal chemistry. The emergence of these entirely new industries was due in some instances to the initiative of individual entrepreneurs and in others to the massive integrated research policy of large enterprises modeled on vertical integration.

Industrialized Physical Chemistry: Synthetic Ammonia

The synthesis of ammonia realized by BASF in 1912 revealed the forceful industrial dynamics engendered by the encounter between organic and physical chemistry, and by the magnified effect of research in strategic activity, both economic and military. This innovation gave a response to the foreseeable exhaustion of Chilean deposits of nitrogen, which threatened to cause a global famine. In the 1890s several intellectuals undertook research in order to find an answer, and two methods entered into competition: a process using an electric arc, and another founded on cyanamide produced from calcium carbide. The latter was perfected by Adolph Franck and Nikodem Caro for the Degussa firm. Much less costly than the alternative, it became the main source of synthetic nitrogen on the eve of World War I. Yet the future belonged to a third procedure, developed by Fritz Haber for BASF, that consumed far less energy.

Judging from patent applications, BASF's research activities after 1900 were progressively oriented toward physical and organic chemistry, though dyes were not ignored. From the beginning of the century BASF was interested in producing nitrogen. In 1900 Ostwald requested that the company acquire a patent he had obtained to create synthetic ammonia. His idea was to combine nitrogen and hydrogen by subjecting them to very elevated temperature and pressure and using a ferruginous catalyzer. The experiment was not convincing. Carl Bosch, a young chemist employed by BASF in 1899, showed that its failure was due to residues of ammonia contained in catalytic iron. This former assistant of one of Ostwald's rivals thereby secured a fine career in the firm, becoming its chairman of the board of directors in 1919. In the short run his project was abandoned. In 1903 BASF became interested in a process conceived by the Norwegian Kristian Olaf Birkeland, which was based on the oxidation of atmospheric nitrogen through the use of a high-voltage electric arc, but negotiations with the Norwegians collapsed. It appeared that completing the project would require considerable financial and human resources not yet possessed by BASF. Discussions with the Norwegians resumed in 1905. With the aid of an engineer, one of BASF's chemists, Otto Schönherr, devised an oven that was tested on a small scale in Norway. A factory employing this procedure on a large scale was constructed in 1907, but its functioning was defective.

That same year, despite some skepticism about the use of elevated temperatures and pressures needed for the catalytic process, the directors of BASF resumed the project. They contacted Haber, who, having been trained by Bunsen at the Technische Hochschule of Berlin, was teaching organic and physical chemistry in Karlsruhe. He also worked in his father's business and was thus located at the frontier between science and technology. Like many other chemists of the time, such as the Frenchman Henry Louis Le Chatelier, he had already scouted the possibilities offered by catalysis and concluded that its industrial application was impossible because of the exceedingly high temperatures it required. He thereupon turned to the use of an electric arc, for which Nernst had demonstrated that a synthesis under very high pressure was feasible. Haber signed a contract with BASF in 1908, and in 1909 he created catalysis at an atmospheric pressure of 175 cm of mercury and a temperature of 550 degrees centigrade by using osmium as a catalyst. BASF soon concluded research on the arc process, but it was still necessary to move from the laboratory to industrial production. Haber himself doubted that it was possible. To make the attempt, the directors of BASF called on Bosch, who had a triple technological culture: that of a metallurgist, a chemist, and a mechanic. He shared "the confidence of an engineer in the utility of metals and machines with the faith of a chemist in the possibility of synthesizing all natural products."[25] He needed to resolve the problems posed by the effect of size: one in particular was the explosion of vats that occurred beyond a certain time of reaction. Industrial production began in 1912 at a pilot factory constructed at Oppau. BASF then constructed a veritable fortress of patents to protect its absolute control of the process.

This research was essentially a multidisciplinary exercise based on science acquired directly from physical chemistry and metallurgy. It was also dependent on the knowledge and skills of chemical engineers through a close association between chemical theory and a nascent chemical engineering corps. It mobilized considerable financial, technical, and human means on a massive scale. Finally, it proceeded according to a systematic logic, to use the expression Baeyer employed in 1880 to describe this scientific research. It can be called planned research tending to eliminate chance. The synthesis of ammonia thereupon demonstrated the utility and efficiency of industrial research laboratories integrated within business firms. This showed moreover that, as had been the case in organic chemistry, new technologies could "emerge from academic research in physical and inorganic chemistry," and that a symbiotic relationship between academic and industrial research might have as much success in those two disciplines as in organic chemistry. "For the first time we see accomplished research put into place, conducted by an interdisciplinary team with a risky strategy and a heavy investment of capital." Furthermore, the results obtained represented a major advance in the history of physical chemistry and of the most theoretical kinetic chemistry.[26]

The Polymer Industry Before 1900

By the end of the nineteenth century, the chemistry of dyes rested on a solid scientific foundation: the analysis of molecular structure and atomic particles, and the theory of valence. This chemistry essentially dealt with materials that could be crystallized or distilled. Always following the same procedures, research work was relatively simple. After the turn of the century, when seventy-year-old Adolf von Baeyer stated that "the field of organic chemistry is exhausted," the course of affairs markedly changed.[27] Certain chemists devoted increasingly sustained interest to organic substances with colloidal properties resistant to crystallization, such as proteins, enzymes, starches, rubber, and cellulose. This interest was stirred by the fact that these substances were increasingly employed in the industrial process, and by the inferences that this problem held for biochemical phenomena. Actually, "the polymers were considered undesirable products: syrups impossible to crystallize or solids impossible to melt, they bothered laboratory chemists who deigned to mention them in notes as 'substances of an unknown nature.'"[28]

These uncertainties of scientific theory did not prevent the development of an important industry based on the utilization of natural polymers, principally rubber and cellulose, and the use of these products blossomed after the middle of the century. In 1839 Charles Goodyear introduced the procedure of vulcanization, which made it possible to produce stabler, harder, more elastic rubber that was more useful in numerous items like straps, telegraph cables, wagon brake shoes, and decorations on chairs or footwear. Empirical learning about rubber was underway, concerning both products and procedures. Increasingly more sophisticated and formalized, it gradually formed a coherent system of knowledge, explained in treatises and including two clearly identifiable channels: basic technology "that permitted common goods to be easily reproduced," and technology oriented toward particular procedures and products.[29] Further inventions followed the appearance of vulcanization. New procedures worthy of mention include the use after 1860 of rubber presses and the mechanization of treatments like lamination. As for new products, there were, for example, the manufacture of rubber coating that satisfactorily replaced leather, the development of products combining rubber with other materials, and the application of rubber to new objects such as toys.

The publication of journals exclusively devoted to the rubber industry, at London in 1884 and New York in 1889, signaled the formation of an international and increasingly autonomous professional community. By this time the use of rubber for electrical lines was already largely developed, and the bicycle and the automobile would open still more considerable outlets. Rubber became a strategic product, militarily and economically. Worldwide consumption grew from 49,000 tons in 1900 to 155,000 tons in 1915. Rubber plantations created in Southeast Asia in the 1890s began full production after 1900. It was no accident that this was the moment when the molecular structure of rubber and the production of synthetic rubber became chosen targets of academic chemical research.

The chemistry of cellulose had its origins in a distant past. It was introduced in the nineteenth century through experimental research on substitutions for natural products considered defective or increasingly rare and expensive. By the 1850s, several university chemists had found the chemical formula of cellulose and defined its properties. Its use in the paper industry then led to unprecedented production from vegetable products and soon from wood. The discovery in 1845 of the explosive properties of a cotton and nitric acid mixture opened the way to a large number of nitrocelluloses and the development of a huge market for explosives. Some of those products could be dissolved in a mixture of alcohol and ether, which transformed them into a fluid called collodion that found various uses, first in medicine and surgery, then in photography in the 1850s. These diverse usages did not, however, provide an solution to the problems posed by viscosity, one of the mysteries that the chemistry of colloids sought to solve. This chemistry extended into artificial fibers and plastics, a word coined by a naturalized American chemist from Belgium, Leo Hendrik Baekeland.

Plastics made their appearance in the 1860s, after a professor at the University of Basel, Christian Friedrich Schönbein, had in 1846 discovered a substance obtained from the reaction between paper and nitric or sulfuric acid, cellulose nitrate. It could be utilized as an explosive but also to produce an easily molded material. An English metallurgist, Alexander Parkes, learned of this discovery and inaugurated a series of experiments using different gelatins and solvents. The resulting substance, called parkesine, was presented at the London World's Exposition in 1862. Parkes initially concentrated on luxury items. He changed his strategy in 1866 by creating new products from nitrates and oils that had much wider use, but several of them proved to be defective, and his enterprise went into bankruptcy in 1868. It was resuscitated by Daniel Spill, who produced artificial white ivory. Albeit without chemistry, the story goes, the American John Wesley Hyatt responded to an offer made by a firm that manufactured ivory billiard balls to anyone who invented a material to replace them. He perfected such a product and, like Parkes, conducted a number of experiments to improve it and to find new uses for it. He was the first to discover the essential fact "that a solution of camphor and ethanol was not only an ideal gelatin but also the best solvent to make cellulose nitrate into a marketable product."[30] Camphor was used for this purpose until the 1970s. Celluloid enjoyed immediate success, and its uses multiplied: detachable collars, shirt sleeves, ornaments, and false teeth (admirably replacing rubber). In 1878 Hyatt patented a technique for casting that allowed belt buckles to be covered with celluloid. Cellulose nitrate thereby became the first thermo-hardening plastic, that is, one capable under heat and pressure of assuming a particular form and keeping it indefinitely. The same product was used for photography and made possible the development of the cinema. Sheets of celluloid made their appearance in 1888. Several inventors like Hannibal Goodwin, George Eastman, and the brothers Auguste and Louis Lumière simultaneously conceived the idea. Goodwin's patent was the best protected, and his product the most widely deployed. Cellulose nitrate was dissolved in an aromatic solution with camphor and other substances of the same type. After drying, an emulsion was applied and a plastic sheet was cut. This product was thus employed in photography but remained too flammable to be used by amateurs.

In the 1880s Louis-Marie Hilaire de Chardonnet had the idea of obtaining artificial silk from collodion. To produce it, he had to conduct countless strictly empirical experiments, both physical and chemical. He thereby managed to derive a silky fiber, but it was flammable and very expensive. Despite these flaws, this product found a rich market before 1910. French artificial silk was then challenged by another cellulose product, viscose, for which the Englishmen Charles F. Cross and his associate Edward J. Bevan secured a patent in 1892. The former had received training at universities in Zurich and London. After working for a firm specializing in linen and burlap, in 1885 he and Bevan founded a center for chemical analysis and consultation. Their patent required a treatment of alkaline cellulose with caustic and sulfuric carbide, producing a liquid called viscose. After aging, it was pressed out and spun. The cost of production was very high, however, and fifteen years passed before it began to compete with Chardonnet's artificial silk at a time when cellulose, increasingly produced from wood, became less expensive. In 1904 the firm of Courtaulds Ltd., interested in alkaline cellulose, purchased the rights to Cross's process and pursued a program of research. Between the two world wars the procedures for coagulating and stretching threads were much improved, and in 1935 the firm marketed threads of very high tensile strength under the name of tenasco.

A third family of cellulose fibers—cellulose triacetates and acetates—appeared about the same time. In 1865 the French chemist Paul Schutzenberger showed that cellulose could be transformed into acetates. Cross and Bevan succeeded in preparing it in 1894. Their discovery was not produced in Germany, even in small quantities, until 1902. Its industrial manufacture posed insurmountable technical problems, and the primary materials used, especially acetic acid, were scarce and expensive. Hence it was not really exploited until 1954, after a more effective solvent of methylene was perfected. In 1904, the Englishman George W. Miles discovered a means to "de-acetize" the product of Cross and Bevan to create a cellulose acetate with solid fibers and films of good quality. This process was adopted in Switzerland by the brothers Camille and Henry Dreyfus, who succeeded in making the film fireproof. During World War I this product was used to coat airplane wings with an impermeable and fire-resistant varnish. The process was further improved, and after the war its production was converted into plastic materials like rhodoid in France and into celluloid films. The manufacture of acetic acid was industrialized when the Dreyfus brothers created the Celanese Company, which produced acetate rayon, much improved by the 1920s. Cellulose acetate became the premier thermoplastic after World War II.

These few examples demonstrate the richness of cellulose and rubber chemistry, which came to define formulas and procedures more precisely by experimental and deductive methods. They spurred the understanding of the properties of substances and of concrete processes to treat them, thereby complementing the preoccupations of physical chemists. This relationship with chemical science might seem essential insofar as these inventors gathered observations, opened new avenues, and posed questions that nourished theorists' reflection and oriented their research. But a contrary thesis is that chemists researching cellulose had a tendency to restrict themselves to a narrow context of resolving precise and particular problems of their clan. In the realm of celluloid, according to Wiebe E. Bijker, most of these inventors made "no efforts to say anything about the structure of the condensation product nor about the chemical reaction in detail." They concentrated all their attention to the discovery of a solvent to replace camphor. "This placed the role of a solvent in a central position in the celluloid technological frame," and "through the patent litigation trials the choice of the right solvent had acquired the meaning of a crucial step in the invention of celluloid."[31] These specialists were totally incapable of perceiving the advantages of the product invented by Baekeland, Bakelite, because they were unable, according to Baekeland himself, "to divorce themselves from the routine of older methods." They were men who "were not engaged in plastic before."[32]

Without contesting Bijker's conclusions, it should nonetheless be recognized that the knowledge these inventors developed was not completely disqualified by the innovation of Baekeland and his successors. They opened new paths. Although for a long while ignored or neglected by leaders of the discipline before being rediscovered, they posed questions and faced difficulties whose resolution provoked a revision of scientific theory. The quantity of experiments and hypotheses they gathered fed the thinking of those who subsequently attempted to formulate a descriptive theory of the structure of substances and to explain their properties and reactions.

The Science of Polymers

Academic chemists did not altogether neglect substances considered undesirable. Despite studies in 1861 by the Scottish chemist Thomas Graham, considered the father of colloid chemistry, research before 1900 on colloids was limited to random observations and remained an esoteric subject. Investigations centered on the classification and nomenclature of scattered systems and on the mechanical properties of colloids. Thus, a mass of data was collected on diffusion, dialysis, filtration, precipitation, electrophoresis, and viscosity. But the chemistry of colloids did not really thrive until after the turn of the century. Then it found a positive echo among biochemists, who benefited from progress made with instruments of observation like the ultramicroscope, invented in 1903. Its acceptance was a response to increasing demand from industries utilizing this type of substance for products such as soap, tannins, prints, resins, and proteins that were ubiquitous in daily life.

Wolfgang Ostwald, the son of Wilhelm, became the propagandist of this new science. After founding a specialized journal (*Kolloïd Zeitschrift*) in 1906, he had a great success in the United States during a tour there in 1900. He made a convert to colloid science of Wilder D. Bancroft, who took up this "chemistry of daily life," as he said, and published a book on it in the 1920s.[33] But though the new discipline had found its Ostwald, it never found either its Arrhenius or its Van't Hoff, and it was less successful in Germany than in the United States.[34] From the outset, colloid chemistry entered a dead end. Its defenders wanted to show that this science was different from that practiced by the majority of German chemists based on analysis of molecular structure. The former argued that colloids were a physical state of interconnected matter forming a complex or aggregate of a large number of small molecules held together by intermolecular forces. These aggregates did not behave like simple molecules. Their mixtures were not tantamount to chemical reactions or to connections in the form of valence. It was only a matter of partial valences, and they were the result of simple phenomena of absorption and action of electrostatic energy.

To account for the nature of these substances, another hypothesis compatible with the theory of molecular structure was possible: that of giant molecules formed by the union between two or more molecules of the same size, which Berthelot had already suggested in 1866. In the 1880s the demonstration of very high molecular weights by several chemists, among them François-Marie Raoult in Grenoble in 1882, did not suffice to gain acceptance of the concept of giant molecules. Before World War I, Emil Hermann Fischer, a 1902 Nobel laureate in chemistry whose work on sugars and the structure of proteins was enduringly authoritative, refused to consider proteins as polymers and rejected the idea that a molecular weight could exceed 5,000. Thereafter, support for the theory of aggregates grew. It was applied to the study of many substances utilized by industry, such as cellulose, starch, protein, and rubber. Carl Dietrich Harries, who devoted intense research to rubber, maintained that it was constituted by a cyclical structure in which aggregate forces held molecules together in a spiral. In the 1920s, this theory of aggregates seemed to be confirmed by the study of polymeric structures with the aid of x-rays. This represented a first step in the unification of the theory of polymeric structures before the emergence and triumph of macromolecular theory.

The discovery of Bakelite in 1912 opened another possible future. Totally different from existing plastics, this product was not the result of transforming a natural substance but was first obtained from a purely chemical combination of carbolic acid and formaldehyde. This process of condensation had been invented by Baeyer in 1872 and by Arthur Michael in the 1880s. Yet neither was much interested in it, since one was studying dyes and the other natural resins. In the 1890s, several chemists like Werner Kleeberg conducted research on the process but abandoned it in favor of "repulsive resin," to use the expression of Leonhard Lederer in 1894.[35]

As noted, several inventors of cellulose products were uninterested in this process because their attention was concentrated on the search for a more effective and less expensive solvent than camphor. Originally Belgian, named professor at age twenty-six and then transferred to the United States as a consultant, Baekeland made a fortune thanks to the sale of the rights to a paper used in printing photographs, velox, to Eastman Kodak in 1899. He had carefully followed the writings on chemistry and knew in particular about Kleeberg's work. Baekeland copied Kleeberg's experiments, plus those of some other chemists, and after five years of research succeeded in mastering the process. He submitted nearly a hundred patents. The most famous of them, called "heat and pressure," was submitted in 1907, but the most important dated from 1909. These patents concerned the production of an entirely new synthetic product from carbolic acid and formaldehyde. With a form that could not be altered, it was called thermo-hardening. This was not a question of manipulating well-known molecules, as in the simultaneous discovery of new dyes, nor an application of scientific theories, but the empirical study of a clearly defined condensation. Its success was obviously the result of a method associated with physical and engineering chemistry. Baekeland did not explain the theoretical basis of his invention before 1913, and it was not until the 1920s that macromolecular theory offered a satisfactory scientific interpretation. The General Bakelite Company was founded in the United States in

1910. In 1926, Baekeland joined the Damard Lacquer Company, an English firm created by James Swinburne in 1910, which manufactured a product very similar to Bakelite, for which Swinburne had submitted a patent in 1907, one day after Baekeland had submitted his own. Swinburne was an eminent electrician who had been president of the British Association of Electrical Engineers. His research, begun in 1902, attempted to resolve the problem of insulation, which in his view was the most serious of the questions posed by electricity. Despite the appearance of many competing products at the end of the 1920s, the market for Bakelite grew rapidly thanks to the implementation of a coherent marketing strategy by Baekeland and his partners. Baekeland thus acquired a favorable position in the electrical and automotive industries, and also in the production of countless domestic objects such as adhesives and plasters. At the time of his death in 1944, annual global production of carbolic resins had reached 125,000 tons.

These various trajectories allow a definition of the techno-scientific elements in the realm of chemistry during the second industrial revolution. Two factions were in competition: those who tended to assert the autonomy of new disciplines, and inversely those who sought to integrate them. In physical chemistry a major bifurcation of scientific theory thus emerged, marked by the explosion of a new discipline that was both radical in its propositions and dominant in its strategy. The rise of this new discipline was closely linked to the development of new industrial sectors like electrochemistry and to a complete transformation of a still largely empirical knowledge devoted to old industries such as metallurgy. In the most recent industries, such as polymers, scientists took care to integrate their theoretical models with the empirical and experimental learning already employed in industrial practice. The discovery of Bakelite opened an unexplored pathway while still imitating the model of artificial dyes in the 1860s and 1870s. Combining the empirical approach of the chemistry of plastics with that of electrochemical science, it arrived at the creation of an entirely artificial product like alizarin at the end of the 1860s.

Business enterprises remained at the core of forming knowledge, for which an entrepreneurial model was perfectly apt. Only its nature was modified. University scientists and engineers with a partially or totally scientific education increasingly participated in the innovation of new industries (electrochemistry, plastics) and on research and development teams within existing enterprises, large and small. In the very large industrial firms, finally, a model of vertical integration determined research activities, for which synthetic ammonia brilliantly illustrated its effectiveness.

Macromolecular Chemistry and Vertical Integration

German Controversies: Mark and Staudinger

Hermann Staudinger had pursued the study of botany at the Technische Hochschule in Darmstadt before joining the institute of Adolf von Baeyer in Munich. He then moved to Strasbourg as the assistant of Professor Friedrich Karl Johannes Thiele, author in 1899 of a theory of partial valence that contributed to the process of aggregation. In 1905, Staudinger finished a thesis on highly reactive compounds called ketenes. He was named to a chair of chemistry in Zurich in 1912. His attention after 1910 was focused on controversies that roiled chemists of rubber, production of which was undergoing an enormous expansion. One of those quarrels was between Carl Dietrich Harries and Fritz Hofmann, both employed by Bayer. In 1910, in a paper concerning "the composition and synthesis of rubber," Hofmann contested the views of Harries and adopted the standpoint of the Englishman Samuel Pikles, who argued that the polymerization of isoprene was a purely chemical process. Hoping to extend this notion to all polymers by launching a research program, in 1917 Staudinger published an article in which he proposed formulas defining the structure of rubber, both natural and synthetic, citing the classical theory of Kekulé. He repeated this thesis in 1920 in a book on polymerization in which he disputed the utility of resorting to a theory of secondary valence forces. In reality this was an intuitive postulate without any experimental basis. In 1922 an experiment with hydrogenation by Jakob Fritschi produced a colloidal substance that was not formed by small molecules held together by partial valences but by giant molecules. Staudinger was able to conclude in 1924 that "the colloidal properties were determined by the structure and size of molecules."[36]

The experimental and institutional framework of this research was radically altered in the 1920s. Research programs based on scientific debates became increasingly integrated into a unified system of theoretical data regarding the nature and structure of polymers as well as the properties of different products like natural or artificial fibers, cellulose, starch, proteins, rubber, and plastics. The main objective of this research was to establish the precise relationship between properties and structure. This obviously practical concern seeped everywhere into technological and scientific knowledge.

The experimental capacity of researchers changed in kind, in terms of both the greater expense of the installations provided for some of them and their technological sophistication. The ultracentrifuge designed by the Swede Theodor Svedberg, Nobel laureate in 1926, was one of those prized by chemical laboratories, but also one of the most expensive. It could attain centrifugal speeds a million times faster than that of gravity. But the greatest novelty doubtless came with the development of crystallographic analysis by x-rays, discovered in 1912 by Max von Laue.

The career of Herman Francis Mark illustrated the role of such instruments in the construction of knowledge. After his entrance in 1921 to the Kaiser-Wilhelm-Gesellschaft Institut, where Fritz Haber had founded a research department for textile fibers, he was assigned under the direction of Michael Polanyi to study the possibilities that x-rays and the ultramicroscope offered for evaluating the mechanical properties of products and exploring "how macromolecules react to changes in their physical and chemical environment."[37] He also worked on rubber, reaching conclusions similar to those of Staudinger. He thereby confirmed that natural fibers were organic substances formed in long molecular chains having a weight of about 100,000 newtons. Some of them had a regular construction and could be crystallized. Mark thus became one of the most eminent specialists of crystallography.

Between 1926 and 1935 this conceptual and experimental framework was the locus of debates that solidified the scientific foundation of polymer chemistry and established a firm basis for the plastics and synthetic textiles industries that became so enormous after World War II. In 1926, at the beginning of his appointment as professor at the University of Freiburg, Staudinger presented his theory of macromolecules. Haber organized a meeting of the German society of physicists to debate the problems arising from macromolecular structures. Partisans of the theory of aggregates dominated the debates, but Staudinger maintained his position, relying on his observations concerning the viscosity of polymer solutions, whereby he had deduced a law of viscosity concerning the relationship between molecular weight and polymeric concentration. He asserted: "Despite the large number of organic substances already known today, we are only at the beginning of a chemistry of truly organic compounds."[38] Opposition to his views was not long in coming from three distinct groups of scientists. The first was composed of traditional organic chemists, among them a half-dozen Nobel laureates for whom the idea of the existence of molecules with a weight of several hundreds of thousands of newtons was inadmissible. The second included such intellectuals as Jean Perrin, Wilhelm Ostwald, and Svedberg, who, while admitting that there were chemical objects of very great size, held that their aggregation or conglomeration through covalent connections was not proof that they were created through the action of "various associative forces—polar groups, complex formations, and others"; hence Staudinger's theory was unnecessary.[39] The third cluster was made up of crystallographers. They contended that because the diminutive size was established for the elementary cell of substances such as cellulose, silk, rubber, and for chemical substances in relation to them, the constituent theory of them could only be small.

Thus began an experimental period in 1926, during which Staudinger's position was weakened by the denial of his request to purchase Svedberg's ultracentrifuge.[40] In January 1927 I.G. Farben recruited Mark to work in a central research laboratory created by BASF and directed after 1920 by Kurt Heinrich Meyer, who had worked with Baeyer in Munich. Meyer recruited only young chemists who had practical training in physical chemistry or physics. He equipped the laboratory with an x-ray machine to study the structure of cellulose and rubber, a machine of electronic diffraction able to examine the structure of simple organic substances, to measure viscosity and osmotic pressure, and to study the structure of polymers. The laboratory also boasted a technical office that could synthesize polymers and test their properties. A specialized research group conducted experiments on the catalytic reaction and synthesis of rubber. Mark participated in these laboratory activities between the publication of academic papers and research, including "close cooperation with existing production lines and with future potential developments." The laboratory was essentially devoted to examination and testing of viscous fibers and acetates produced by the company and to synthesis and evaluation of new polymers.[41] Mark and Meyer achieved several remarkable successes, and in 1930 BASF became the first manufacturer of polystyrene.

Parallel to this practical research, BASF pursued theoretical study of natural polymers and used x-rays to investigate the arrangement of atoms in long chains. Before 1927 Mark had not excluded the presence of large molecules, though he conceded that their existence was unproven. According to him, Staudinger had reasoned by simple extrapolation from a study of the properties of long series of oligomers, showing that chemical modifications of polymer chains did not change their colloidal properties. Mark and Meyer later approached Staudinger's standpoint, proposing in 1930 a compromise between his theory and the theory of aggregates by suggesting the existence of a micellar weight. Thus, they asserted that cellulose was composed of micelles, that is, aggregates formed of "primary valence chains (*Hauptvalenzketten*), or long-chain molecules held together by 'special micellar forces' (*besondere Micellarkräfte*)."[42] Staudinger rejected this compromise, although he admitted the existence of intermolecular energy. In fact, the theory of micelles offered a unified system for representing all macromolecules. Partisans of the theory of aggregates rallied to the thesis of Mark and Meyer, and by doing so they indirectly supported macromolecular theory.

After 1926, Staudinger redirected his research toward the study of proteins and the phenomena of viscosity. He demonstrated that molecules longer than cellular crystallographic units could exist. Vigorous resistance to Staudinger's ideas persisted until the end of the 1920s, but a movement in support of macromolecular structure gradually prevailed, including cystallographers like J. R. Katz, who began to doubt that polymeric molecules were smaller than elementary cells of crystal. A consensus formed in favor of a compromise in the 1930s, thanks to the work of Mark in Germany and especially Wallace Hume Carothers in the United States.

Staudinger's definition of the concept of macromolecules provided scope for further development. For him macromolecular substances were multimolecular, that is, composed of molecules of variable weight and size. This variety of configurations explained the great diversity of their properties, such as fiber quality, elasticity, flexibility, viscosity, and distention. He classified them into two groups, linear or spherical, depending on their structure. This, he wrote, "affects the physical and chemical properties of the substances considerably more strongly than is the case with the low molecular compounds." For Staudinger, macromolecules opened a new chapter in organic chemistry, because "it is understandable that new properties will therefore be found which are not possible in low molecular materials." Thus, the foundations of classical organic chemistry were sustained.[43]

In sum, the role played by these controversies among laboratories and technical centers, whether at universities or in enterprises, marked a fundamental stage in the development of techno-scientific knowledge. These disputes had become the laws of contacts between theories and experiments, science and industry. The laboratories of enterprises did not have the same ambition or organization as the university laboratories. Yet they followed the same conceptual paths, participated in the same controversies, and utilized the same instruments. Hence these laboratories essentially encouraged the convergence between these two types of research.

Carothers

In the United States, the chemistry of colloids enjoyed great success in the first quarter of the twentieth century. It was taught in particular by Wilder D. Bancroft and James B. Conant. At the beginning of the 1920s, the controversy about the real nature of polymers was as intense in America as in Germany. Conant adopted the theory of aggregates, as did many American scientists. Wallace H. Carothers, however, then assistant professor at Harvard, leaned toward the theory of macromolecules. This young chemist's career took a new turn in 1927, when Charles Stine, the research director of DuPont, persuaded the executive committee to launch a program for fundamental research on polymers. Stine justified this step by showing that DuPont's various chemical activities depended on common scientific knowledge that industries and universities were not actually exploring. The program's goal was "to establish or discover new scientific facts" and not, as hitherto, "to apply established scientific facts to practical problems." Stine thought university research had been deficient in numerous realms, so that "applied research is facing a shortage of its principal raw materials."[44] He showed that the research policy he advocated, oriented toward fundamental research, had been adopted by the German chemical industry and especially by General Electric in the United States, whose laboratory, directed by the future Nobel laureate Irving Langmuir, had been created in 1900. The idea that fundamental research could resolve practical issues had also been clearly implemented after 1910 in several enterprises such as Eastman Kodak. Stine made a distinction between pure and fundamental scientific research, which "might result in something of great value or might come to naught. But the latter is bound to result in the discovery of new highly useful and in some cases indispensable knowledge," and thus "pioneering applied research is a form of gambling."[45]

Stine proposed paths of research for DuPont to follow: colloid chemistry, catalysis, creation of physical and chemical data, organic synthesis, and polymerization. He noted in regard to catalysis that its action was unknown, and in regard to polymers that almost nothing was known "about the actual mechanism of change which takes place, so that the methods are based almost solely on experience." To realize this program he recruited "men of exceptional scientific promise but [with] no established reputation."[46] One of them was Wallace Carothers. Employed in 1928 by a group of organic chemists in a laboratory of pure science, he was assisted by two specialists in colloidal chemistry: one a student of Svedberg, Elmer O. Kraemer, and the other a student of Bancroft, James B. Nichols, who worked on the relationship between the size of particles and the properties of pigment. When he entered the DuPont firm, Carothers had already rejected aggregate theory and adopted a view close to that of Staudinger. He believed polymers were formed of molecules like other chemical substance and were comparable in that regard to cellulose. He proposed an ambitious research program intended to furnish general indications about the physical properties of polymers and to define an experimental approach based on synthesis, inspired by the work of Emil Fischer. For this purpose Nichols constructed an ultracentrifuge similar to Svedberg's. According to Yasu Furukawa, Carothers "combined small molecules one by one through well established organic reactions" and "definitively obtained long molecular chains of known composition."[47] In 1929 he presented a general theory of the condensation of polymers by distinguishing between two polymeric types: one, studied by Staudinger, derived by the addition of monomers; and the other formed by a reaction of thermo-molecular condensation. He defined polymerization as "the chemical union of many similar molecules either without or with the elimination of simpler molecules." He placed rubber in the first category, silk and cellulose in the second.[48]

When he presented his program to Stine in 1928, Carothers asserted that "commercially important resins" probably belonged among the chemical substances he hoped to study.[49] As it happened, the industrial fallout of his research was soon to come. Synthetic neoprene was accidentally discovered in the course of a project intended to analyze a rare chemical compound formed of three molecules of acetylene, DVA (divinyl acetylene). This finding resulted from collaboration among Carothers, Elmer K. Bolton from the department of dyes, and Arnold M. Collins from the chemical department. Bolton had worked since 1925 on synthetic rubber, attempting to derive it from butadiene, a product obtained from acetylene by using a catalytic method devised by Julius A. Nieuwland at the University of Notre Dame. This research, after being abandoned because of the low cost of rubber, was revived by Collins, whose work on DVA produced, on 17 April 1930, a new compound having the same properties as rubber. It was his department of chemistry that took charge of the program for producing neoprene.

In 1928 Carothers pursued his research on the properties of long molecular chains in close cooperation with the MIT graduate Julian Werner Hill. He used a molecular still that allowed a displacement of the equilibrium of polymeric condensation to a much higher level of reaction and to a very great weight. Carothers and Hill thus reached a synthesis of super-polymers having a weight of more than 5,000 newtons. By experimenting, they discovered at the same time that the properties of super-polymers did not depend solely on their chemical structure but also on their physical treatment, especially stretching. This permitted "long molecular chains [to be] arranged in a highly ordered array parallel with one another."[50] Carothers and Hill thereby laid the theoretical foundation of the production of completely synthetic fibers. In September 1931 Carothers proudly announced to the American Society of Chemistry that he and Hill had "clearly demonstrated for the first time the possibility of obtaining useful fibers from strictly synthetic materials."[51] In the short term, he thus created a melted polymer in the form of a very solid and supple fiber with a molecular weight of 12,000 newtons. But as these fibers proved to be too soluble, the product was still only a promise.

Named director of the chemistry department at DuPont in 1930, Bolton believed the research of his firm should be limited to solving specific, precisely defined problems. In 1932, when he turned laboratory research in pure science toward solely technological goals, several researchers in his department resigned. The academic era was terminated.[52] Carothers disagreed with this orientation and declared that fundamental research should be guided by scientific and not commercial considerations, adding that he regretted having accepted this work in 1928. He continued his research on polymerization by studying the effects of molecular geometry. His team conducted experiments on various super-polymers, but their textile properties remained deficient. In 1934 he directed his research elsewhere by using a purified ester extracted from amino acid.

In May 1934 Donald Coffin succeeded by this means in producing a synthetic fiber with satisfactory properties, followed by many others from which Bolton selected the one having "the best balance of properties and manufacturing cost of the polyamides then known."[53] Nylon produced from benzene was born. It was also in 1934 that Carothers entered a depression that ended with his suicide in 1937. The tension he felt between the desire to do fundamental research and the obligation to pursue commercial research was doubtless relevant to his condition. Research to find a marketable fiber was conducted by DuPont's rayon department. From the outset, its chief had believed in the future of totally synthetic fibers and had already participated in the work of Carothers's team. This evolving research required five years of effort. Nearly everything had to be invented, and excepting neoprene, no synthetic product existing on the market required as tight a control of polymerization as nylon. Attempts to transfer laboratory procedures to an industrial scale failed. The researchers turned to a process of emulsion patented by I.G. Farben. It was necessary to design a program of basic research combining knowledge from physical chemistry and chemical engineering. From the former, they borrowed procedures of high pressure and catalysis of synthetic ammonia; the rest was supplied by the department of rayon. DuPont could not have exploited the process without this know-how. But all went well in the end, since the Carothers version of the nature of polymerization through condensation afforded the developers of nylon and neoprene the intellectual tools necessary to produce commercial nylon.

Before 1936 Carothers published a series of studies on the application of his method to a large number of substances like polyesters and polyamides. These texts founded the new chemistry of polymers. After pondering the writings of Staudinger, Meyer, and Mark, he offered a clear definition of terms and a coherent classification of substances as well as a general interpretation based on the concept of macromolecules. Carothers rejected Meyer and Mark's theory of micelles, which he thought were nothing but micromolecules. He likewise rejected the analytic method Staudinger had used to demonstrate his theory, and Carothers criticized Staudinger's dogmatism. Carothers was convinced that only a synthetic method could provide a solid basis for theory. Whereas Staudinger had been content to study the final product, Carothers was interested in the process. It is in this sense that Carothers may be considered a physical chemist. American chemists' support for the macromolecular thesis was spontaneous. It was not Staudinger but Carothers who prevailed.

Founding the Physical Chemistry of Macromolecules

Appointed professor in Vienna in 1932 after his stint at I.G. Farben, Herman Mark started a program of research explicitly intended to construct a "physical chemistry of polymers." This included the study of polymerization and of polycondensation by extending the work of Carothers and his team, the study of contractions of rubber in relation to problems of entropy, and the study of spiral chains in relation to the problem of viscosity. These various subjects required the use of statistical and mathematical methods, especially when calculating probabilities, and also of concepts borrowed from physics. Conscious of his own limitations in these disciplines, Mark called on a young Viennese mathematician, Eugen Guth, and a young Swiss physicist, Werner Kuhn, a disciple of Karl Freudenberg. Competing with Kuhn, Goth later claimed that he and Mark were the founders of the physics and physical chemistry of polymers. Among other things, they clarified some problems posed by the elasticity of rubber and statistically analyzed flexible chains of molecules. The Vienna laboratory in fact had an interdisciplinary character since, as Mark wrote, "now you start with organic chemistry, you go into physical chemistry, and you end up with physics."[54] An identical orientation appeared at DuPont, where in 1934 Carothers recruited Paul J. Flory, like Kuhn a specialist in photochemistry, and took advantage of his mathematical competence, using it to study the effects of molecular size and viscosity.

Many other research sites besides the University of Vienna and the DuPont laboratory participated in the construction of the physical chemistry of macromolecules. In September 1935 a conference held in Cambridge marked, with few exceptions, the definitive acceptance of macromolecular theory and the rejection of several of Staudinger's positions. His dual vision of a molecular structure composed of rigid elements and of molecular properties governed by viscosity and the effects of size, such as he had defined it in the 1920s, encountered strong opposition. Kuhn and Carothers opened the way to a new interpretation of viscosity and macromolecular structure. Kuhn advanced the concept of flexible spiral chains and asserted that polymers were formed of similar chemical molecular mixtures distributed according to their size. For his part, Flory vigorously challenged Staudinger's theses about the relationship between size and reaction. He introduced the concept of a transfer of activity from one molecule to another in a chain of macromolecules. To describe these phenomena it was possible to employ the calculation of probabilities with the conceptual and statistical tools provided by thermodynamics and kinetics.

From the end of the 1930s to the 1960s the theoretical basis of the physical chemistry of macromolecules gradually took shape. This was the work of physical chemists like Flory, who after leaving DuPont in 1938 became one of the leading theoreticians in the discipline and received a Nobel Prize in 1974, and the Dutchman Peter Joseph W. Debye, Nobel laureate in 1936. The latter immigrated in 1940 to the United States, where he participated in a program of military research on synthetic rubber. Following a paper written by Albert

Einstein in 1910, with the aid of light dispersion, he developed mathematical equations permitting calculation of the molecular weight of polymers.

Birth of the Industry of Synthetic Polymers

In the wake of macromolecular physical chemistry after the early 1930s, a flourishing polymers industry was capable of supplying great quantities of a multitude of totally synthetic and generally useful products. This remarkable takeoff was largely due to a system of knowledge that gave coherence not only to laboratory research and theoretical constructions but also to industrial projects. Staudinger, Mark, and Carothers were only three of the many scientists who contributed to this development. No one who took part in launching these new products overlooked their work or failed to gain inspiration from it.

Three brief examples may suffice. Polyvinyl chloride, or PVC, was one of the many polymers that appeared between the two world wars. This product had been extracted for the first time by Henri Victor Regnault in 1835. In 1912 the German chemist Fritz Klatte, who worked for Hoechst, conducted the initial research to produce an industrially feasible product, and during World War I several tons of PVC were manufactured. In the 1920s Klatte himself and other chemists resumed research. A satisfactory solution was found at I.G. Farben in the early 1920s thanks to a procedure of polymerization by emulsion conceived by Hans Fikentscher, who was researching synthetic rubber. This success was an outcome of the vast experience gathered by I.G. Farben laboratories in the realm of polymers, the mastery of plastics, and the precocious reckoning of early theoretical data from macromolecular science.

Polyethylene was discovered through a laboratory incident in the course of fundamental research undertaken by the director of the English firm ICI, Francis A. Freeth, which concerned not the structure of macromolecules but the use of high pressure while testing the reactions of ethylene in the presence of aniline and benzene. Much like Carothers at DuPont, Freeth was working without specific research goals. His large research team was directed by Eric W. Fawcett, Reginald O. Gibson, and Michael W. Perrin, each of whom investigated a particular aspect of the project. In 1934 a white powder, which proved to be a high polymer created by the linear reaction of ethylene molecules, was discovered following the malfunction of a device for the polymerization of ethylene. Industrial production of it began soon after World War II.

Carothers directed his research in industrial usage toward aromatic rather than aliphatic polyesters. It was an English university-trained chemist, John Whinfield, who opened a breach in the 1940s. After working with Charles Frederick Cross, he joined the Manchester laboratory of the Calico Printers Federation in 1924. At the beginning of his research, which culminated with the discovery of terylene, he possessed a profound understanding of cellulose chemistry acquired from Cross as well as an acquaintance with the work of Carothers. Whinfield's idea was to treat ethylene glycol with teryphthalic acid rather than phthalic acid. The result was immediate. Few discoveries have been conceived from the start with such perfect precision. His finding was confined during the war to ICI, a firm with which Whinfield was under contract through his service in the Ministry of War. He joined ICI in 1947, but production on a large scale did not begin until 1952 with the construction of a huge petrochemical plant at Wilton.

After the war, polymer industries developed within a general framework of large enterprises that after 1950 moved from carbo-chemistry to petrochemistry. The treatment of synthetic products was radically modified by two successive discoveries about the process of polymerization under low pressure through the use of aluminum as a catalyzer, perfected in 1949 by Karl Waldemar Ziegler, Nobel laureate in 1963, during a program specifically devoted to the study of catalysis. At first this produced only low polymers, but in 1953 the treatment was extended to high polymers. In 1954, in collaboration with the Italian Giulio Natta and the firm of Montecatini, Zeigler uncovered a process for producing synthetic prophylene, a homologue of ethylene. Ziegler's process did not compete with procedures of high pressure but led to the production of plastics with various qualities. At the same time, oil companies like Phillips Petroleum and Standard Oil of Indiana started processes of polymerizing ethylene rather similar to those of Ziegler and Natta by tinkering with catalyzers and pressures. Thereby it became possible to create an entire range of materials from ethylene and its homologues by varying the means of polymerization.

Fundamental research on macromolecules, like that developed in the early 1920s from the first work of Staudinger, laid the basis of a synthetic materials industry of global scale and infinite diversity. Not until the 1970s was its pre-eminence partly eclipsed by the emergence of other new materials, also through the industrial application of fundamental research. Advances in basic science during the 1960s and especially the 1980s placed into question the traditional schemas of physical chemistry as a result of experiments conducted with lower or higher temperatures and the utilization of techniques from particle physics. This research permitted the discovery of more new materials such as metallic glass, artificial crystals, and liquid crystals. Altogether, these materials steadily improved in performance, all the more so after the rapid development of compound materials in the 1970s. Specific properties of metallic glass and crystals explain their use in the aerospace and automobile industries. Ceramic technology offered products whose resistance to wear and scratches made them natural candidates to replace metals and mechanical parts. So-called functional ceramics were used largely in electronics and magnetic recordings. Semi-conductors realized unexpected success in speed and miniaturization. Specialty polymers also upset usual practices. Adhesives, for example, tended to replace bolts, rivets, and soldering. Polymers could be used either as insulation or as electric conductors. Glass fibers continually increased their transmission capacity. Compound materials allowed production of parts for practical use and were often employed by the aeronautics, space, and auto industries as well as recreation industries in products such as sails, skis, and tennis rackets.

This evolution of the history of the polymer industry consisted of a dual dynamic: the use and the transformation of knowledge born from the confrontation between theoretical controversies and industrial practices.

Summing Up Chemistry

The construction of knowledge, which has dominated the history of the twentieth century and will continue to rule global learning so long as it adapts to information technology, became evident in organic chemistry in four stages.

1. Until the 1870s the evolution could be defined in several respects. First, organic chemistry gradually attained a clear disciplinary identity around which a hierarchical system of academic learning was organized thanks to the growing professionalism of its research activities. Second, the chemistry of dyes was transformed by merging two cultures: that of colorists, inheritors of a long past of accumulated know-how; and that of university-educated chemists. This fusion was realized through cooperation among many individuals within the framework of easily identifiable social networks, and through their mastery of the two cultures. Third, this merger promoted the evolution of laboratory research toward increasingly systematic and programmed methods through theoretical advances that appeared after 1870. Fourth, the principal actors in this transformation were business enterprises, often artisanal in origin, that were created by chemists and colorists whose model was William Henry Perkin. They were integrated into networks of local research, had bountiful accumulated experience, and were nearly all in close contact with academic circles that were themselves deeply involved with those enterprises. They remained in fierce competition, most obvious in court proceedings concerning the legitimacy of patents.

2. Between the 1870s and 1890s the chemistry of dyes was transformed. Certain enterprises, principally German, circumvented the jurisprudence of patents and developed a systematic and coherent research policy. They developed rapidly thanks to the multiplication of new and cheaper products emerging from organic synthesis. This success was owed to a strategy of creating and controlling knowledge that gained general acceptance. That strategy depended on systematic recruitment of engineers emanating from universities and technical schools, but especially on the creation of research departments in large

enterprises. These services had a clear function of keeping the enterprises constantly up to date with innovation and research. This research maintained an essentially industrial orientation, but it was also open to fundamental research either by cooperation with scientific consultants or, on occasion, through particular programs. Such collaboration contributed to modifications of learning in the academic world. The utility of research in chemical theory became self-evident, and its indispensable data were taken into account by industrial research laboratories whose own data in turn had repercussions in fundamental research.

3. Beginning in the 1890s the well-being of organic chemistry, whose practitioners occupied the most prestigious university posts, was gradually shaken by the emergence of competing disciplines. Interdisciplinary competition and the controversies it engendered thus became an important factor in the dynamics of constructing knowledge. But the scientific disciplines had to meet another challenge posed by the irrepressible rise of autonomous learning associated with the technological development of the second industrial revolution: the special needs created by this new technology favored the evolution of new industries like cellulose and rubber, which employed empirical practices of know-how that the dominant chemical science was incapable of understanding or improving. The role of enterprise was also significant in other emerging sectors belonging to physical chemistry, such as electrochemistry and electrometallurgy. Moreover, in both laboratories and factories the construction of new industrial and fundamental knowledge was dependent on the use of increasingly expensive and sophisticated instruments. This process intensified competition among universities and also created a decisive advantage for large industry.

4. The interwar period marked the culmination of the process of theoretical integration, begun in the 1890s, between organic molecular chemistry and physical chemistry in the realm of macromolecules. This merger was only one aspect of progress concerning the entirety of chemistry, whose essential data would not be challenged again until the 1970s. Within the limits of macromolecular chemistry it is clear that the role of controversies was paramount in this process. Some of those involved explicitly sought to reach a consensus, while others assumed a defensive posture. Still others found an experimental exit and thus a way to surmount contradictions and polemics by gathering an immense amount of data and viewing them in perspective. It became increasingly difficult in this process for researchers to avoid a context inextricably uniting industrial practices and theoretical research. Nonetheless, the distinction between theoretical learning and engineering science did not lose its relevance. Chemical engineering more and more asserted its autonomy while becoming, to be sure, more scientific. Meanwhile, the research model adopted by large industry was doubly reinforced. Quantitatively, the financial means

deployed increased considerably, especially in the 1920s, a trend that accelerated during and after World War II. Qualitatively, enterprises were progressively involved in fundamental research. But these strategies had their limits because of uncertainties about this type of inquiry and the exorbitant costs of development. Yet complex systems emerged through the cooperation among business firms, universities, and consultants.

The history of macromolecular theory thus affords a better understanding of the transformation of knowledge during the century between the 1860s and 1950s. In the first place, the increasing role of controversies became a dynamic factor in the academic world. They unleashed a proliferation of competing hypotheses and a surge in experimental approaches. Although their intensity could be aggravated by ideological or disciplinary tensions that sometimes made dialogue and compromise more difficult, compromises always had a great impact. For example, they allowed the dissolution of barriers between classical organic chemistry and physical chemistry, which radically altered the very nature of chemical science. Illustrative in this regard was Herman Mark, who in 1932, after becoming a professor in Vienna, recognized that his training in mathematics was insufficient to conclude certain research and so recruited the Viennese physicist Eugen Guth and Swiss physical chemist Werner Kuhn to augment his investigations.

Secondly, the symbiosis between theoretical research and industrial research appeared complete, even though intellectuals like Mark and Staudinger did not confuse the two disciplines, whose orientation remained intrinsically opposite. Yet one of the most salient facts was doubtless the theoretical nature of some data necessary for the development of industry. The role played by laboratories and centers of technical controversies, whether at universities or enterprises, opened a new era in the construction of techno-scientific knowledge. They became sites of encounter between industry and science. An enterprise laboratory did not have the same objectives or organization as a university laboratory. Yet both followed the same conceptual procedures, participated in the same controversies, and used the same instruments. Both thereby made essential contributions to the convergence of research.

Meanwhile, research in enterprises became increasingly massive. Programmed according to the enterprises' strategic choices, it subtly combined industrial preoccupations and theoretical perspectives. In his autobiography, regarding his activities at an institute of fiber research, Mark wrote: "Science knocked at our door, at first hesitantly, then more and more emphatically." It was, he added, a welcome guest.[55] At I.G. Farben, Mark and Kurt Meyer were important actors in the scientific debate Staudinger had opened over macromolecules, even as they directly participated in the controversial industrial applications of their research. They recruited organic chemists and also chemical engineers to their laboratory. Thus, the organization of gigantic research projects at I.G. Farben tended to retain the model of innovation that characterized European enterprises at the end of the nineteenth century. Its basic principle was the close integration of research within the activity of production. The effect of this research effort depended on the intensity of dialogue between laboratories and the other services of the enterprise, as well as on the quality of dialogue between the enterprise and the university. But in the United States, at DuPont, the organization of research rested on a separation of fundamental research projects from applied research rather than on their integration.

Finally, the success of the model of integrated and massive research did not eliminate the individual entrepreneur as an actor in innovation, as numerous easily cited examples demonstrate. The takeoff of the electrochemical and electrometallurgical industries in the 1890s was the result of initiatives by individuals or small groups—founders of innovative enterprises—integrating into a network of freely circulating knowledge, and possessing technological skills and/or useful scientific learning. One may also list the role in the petroleum industry of a naturalized American, the Frenchman Eugène Houdry, who explored the catalytic cracking of petroleum in the 1920s. This model also applies to the successive waves of innovation that occurred in electronics and the computer industry between 1950 and the 1980s. Researchers swarmed out of universities and large research laboratories to found new business enterprises to develop their ideas and inventions.

In the 1970s the model of integration and gigantism was challenged by economic crisis and by doubts expressed about its efficacy. This crisis was viewed as the result of a slackening in productivity and innovation. Some even claimed that technological change had reached its limits. The vertical, rigid, hierarchical organization of research centers was blamed for hindering individual initiative. To respond, it was thought necessary to adopt a different, decentralized model based on horizontal organization as in Japan. Due in large measure to this self-criticism, new practices of research and administration were to appear during the 1980s.

Part Three

The construction of knowledge described here constitutes only one of the components of technological change. It must be placed into a global perspective that takes into account the ensemble of social, political, and cultural factors created by technological choices and their consequences. Construction of knowledge is a product of both a continuous sharing of information and a permanent confrontation among actors with divergent interests. Hence it is not sufficient to reconstruct the intellectual trajectory of individual sectors, as they should also be located within a global context.

Two guidelines point the way:
1. A dynamic network of technological interdependency is evident in the configuration of products and procedures. Engineers must constantly remedy the insufficiencies and dysfunctions of productive systems and the maladjustments of products to customs and social needs. Progress is therefore limited to certain pathways, determined by initial technological and scientific choices. There is no better illustration of this development than the history of tools by Paul Feller and Fernand Tourret, who confirm that if "tools are subjected to constant variation," it is because "workers do not accept unsatisfactory objects," that is, those poorly suited to the uses for which they were conceived. A tool is created to meet a precise need and evolves with the appearance of new needs. This "constant mutation" proceeds along a trajectory defined by a worker's intention to adapt a given tool to his own needs and to new usages. The entirety of a technological system is subject to the same logic.[1]
2. Technology is a social unit. It is the product of activity by networks and social groups, formal or informal, which attempt in a competitive arena to promote new techniques in order to gain profits, respond to social expectations, or simply fulfill the dreams of engineers. These groups may establish territories and create sources of innovation.

Chapter 8

Technological Interdependence and Consumer Needs

It has been previously demonstrated that technological trajectories evolve according to the malfunctions of a productive system or products' unsuitability to their purpose. Only innovations are capable of supplying an adequate response.

Malfunctions can result from incidents, accidents, or dramatic breakdowns, as well as from persistent operating errors that are poorly dealt with. This was the situation of the technologies of mechanics, energy, and materials throughout the nineteenth century. In each of these sectors, technological uncertainties were incessant and systems were in a permanent state of revision. Meanwhile, they had to adapt to social needs. Appropriate responses depended on initial decisions by champions of a new process or product. Their direction resulted from positive reactions that slowly emerged as the use of technological objects multiplied. These trajectories also involved a process of apprenticeship, illustrating the cumulative character of technological learning. As the number of uses grew, the reactions were amplified and gradually became less individual and more collective. Solutions were sought in large economies and lower costs, but also in a necessary interplay between retained objects and new ones.[1]

Thus, it appears justified to construct a theory of technological change by emphasizing the role consumers played in the emergence of innovations. Such an approach has both a macro-economic and a micro-economic dimension. The former is based on Wassily Leontief's 1941 analysis of exchanges among sectors. The latter takes into account the relationships established among various actors in innovation: between customers and merchants of a product, between inventors and users of procedures, or between one department and another within an enterprise. Leontief describes what he calls the "general interdependence" of a national economy, that is, "the interrelations among the different parts of a national economy."[2] The coefficients he calculates in his tableau of intersectorial exchanges to describe these relationships are exclusively economic in nature, but their value is essentially determined by technological data. Hence they echo the "technological interdependencies" subsequently defined by Nathan Rosenberg.[3] Like Leontief's coefficients, these interdependencies are in perpetual disequilibrium, since no technological development is ever in a state of perfect stability, either because of the imperfect adaptation of objects and procedures to their usages or because of dissimilar timing in their evolution. This permanent inconsistency is revealed in the various forms of dysfunction caused by major technological catastrophes or by the numerous daily incidents of systems, often requiring compromise solutions that can only be provisional. The need to surmount such obstacles is the primary source of technological change and innovation.

Thus, the history of materials is entirely dependent on their adaptation to new deployment of objects and systems that are in perpetual transformation. They are at the core of technological interdependence. The mechanization of manufacturing processes in the first half of the nineteenth century was realizable only by using metals in machinery and changing the basic nature of the metals utilized in production. The democratization of luxury items such as silverware, glass mirrors, and aluminum objects resulted from a massive reduction in the costs of production of these noble and rare materials. The development of aviation rested on the substitution of aluminum for wood in the manufacture of aircraft in the 1920s. Until recently, the unsuitability of materials to their uses was a major impediment to the rise of certain technologies. But producers were under increasingly strong pressure from consumers to furnish products more precise and useful. To respond to these demands, a veritable science of materials was born that considerably enlarged the spectrum of possibilities. The technology of materials has ceased to be a hindrance and has spawned new and unanticipated usages.

Iron Metallurgy in France in the Nineteenth Century

The history of iron and steel manufacture after the twelfth century crossed through four stages: the medieval era up to the fifteenth century; the early modern period from the fifteenth to the eighteenth century; the time of the first industrial revolution from the mid-eighteenth century to the 1840s; and the rest of the nineteenth century after the 1840s. These four epochs of production were greatly conditioned in their development by the particularities of demand and by the slow transformation of usages. The production of metal and its utilization evolved in a complementary manner within the framework of a somewhat chaotic dialogue marked by delays of supply to meet consumer needs and by customers suspicious of novelty.

Medieval Metallurgy: Products, Uses, Techniques

The uses of iron were associated with most technological innovations during the rise of medieval civilization. Archeological discoveries have demonstrated growing use of iron objects in daily life between the sixth and eleventh centuries. A real hunger for iron seized the Occident at the end of the eleventh century, and a European market for metal products developed by the end of the thirteenth century. The increasing use of iron revolutionized agriculture during the eleventh and twelfth centuries. The tools and therewith the gestures of peasants evolved in an often imperceptible manner. Traditional tools—pitchforks, shovels, hoes—and also plows and cart axles were fashioned of iron. Plowing furrows in soil before seeding began in Europe during the tenth century. The harvest was then done with a metal sickle. Windmills and wagons were hooped and horses were shod. To meet these various needs and usages, mines were opened during the late Middle Ages in Catalonia, Poitou, and the Ardennes.

The use of iron in buildings expanded considerably in the thirteenth century. Bolts, locks, and grills were increasingly placed into walls, as were nails, bars, and clasps. Locksmiths and blacksmiths entered the workplace. Wooden fasteners in masonry were replaced by metal, and large metallic pieces appeared in the Gothic architecture of the late thirteenth century. Metal bands were inserted into structures like the upper part of Notre Dame de Paris in the thirteenth century, the Papal Palace at Avignon in the fourteenth, and the dungeon of Vincennes, whose construction used twelve tons of iron and sixteen tons of lead. Such techniques were significant in raising the height of buildings.

In the eleventh and twelfth centuries, warriors' weapons—lances, swords, maces, knives, axes—were increasingly lethal, and the technology of defensive arms was rapidly transformed. It was in the eleventh century that iron mail first appeared, originally made of little metal plaques cut in rows, then of knots forged by hand and riveted together. Heads were covered by a bassinet or helmet. The brigandine, made of leather encasing small metal sheets, had great success in the fourteenth and fifteenth centuries. Shields disappeared from battlefields in the fourteenth century, but they were still used in tournaments. At the same time as cannons, there appeared plated armor formed of fifteen pieces of iron, attached by leather straps, that created a carapace around the body. To meet this demand, armor manufacturing developed in several regions of Europe. The fifteenth-century ironmasters of France and Italy were soon outdone by the artisans of Augsburg. The most elegant offensive weapons were swords and knives with iron blades and steel tips.

The refinement of technology depended on the function of products. Flat objects like ploughs, wheel rims, agricultural tools, domestic items, and mechanical clocks were made of iron. As soon as the process of iron-making became known, pots, cauldrons, cannons, cannonballs, and oven grills were cast, considerably increasing their efficiency. Steel was a rare product, derived from smelting magnesium metals and used mainly for the manufacture of swords and knives.

Iron production slackened in the fourteenth century due to a lack of wood for fuel, but it had a brisk revival after 1450 through building projects and development of artillery. Furnaces were the property of lords, secular or ecclesiastical, except in the peasant communities of Normandy. In that case furnaces were grouped in certain villages where production was overseen by smelters belonging to guilds in which the transmission of their procedures was strictly hereditary. This peasant ironworking experienced a serious crisis in the fourteenth and early fifteenth century but was then reconstituted at the same time as a competing seigniorial iron production.

The technology of ironmasters, inherited from the Romans, utilized a direct method with a single process closely tied to the mines. A radical innovation that spread throughout Europe in the twelfth century was the use of hydrological power to mechanize certain operations like hammering and deploying bellows. Medieval historians have highlighted the role of the Cistercians, an order created in 1098, in development of the hydrological forge. Cistercian monks in Burgundy produced iron to meet their need for agricultural tools. But between 1160 and 1180 they began to sell iron whose excellent quality was widely recognized. The first evidence of use of tilt-hammers driven by waterwheels, at the Cistercian monastery of Clairvaux, has been dated from 1135. In the course of the twelfth and thirteenth centuries, these monks organized a number of operations around this major innovation to create a single hydrological system from the mines to the forge, including gathering wood, washing minerals, and deploying furnaces and bellows. To these ends, they gathered the best specialists in minerals, mechanics, and hydrology. They were doubtless the most innovative users of waterpower, which emerged in Lombardy around 1250. This practice subsequently spread to the Rhine, the Danube, and regions of France in the fourteenth century and to the Pyrenees at the end of the fifteenth.

A second innovation, the process of indirect smelting, appeared about 1450. It was the result of transferring know-how from the production of copper and lead to iron. The first attempts occurred in Germany and Sweden. In this process, watermills were used to activate grinders and furnaces. Complex hydrological systems assured a regular supply of water, while smelting was accomplished with a blast furnace. The product, composed of iron, carbon, and residues, was then refined. This procedure was improved by Walloon ironmongers, who had learned from German metalworkers imported in the mid-fourteenth century by Marquis Guillaume de Namur. This method allowed a combination of hammering and refining with the capacity of blast furnaces, making mass production possible. The process spread to Lombardy, Burgundy, and the Paris region. It often necessitated the building of complex hydrological systems with canals and sluices to assure a regular flow of water. Thereby a new material was created: cast iron, which could be molded like bronze but was fragile.

The Early Modern Period

Between 1450 and 1880, the use of iron and cast iron developed at first rather slowly in Western societies, then more rapidly after the mid-eighteenth century, penetrating all agricultural and industrial activities as well as domestic consumption. It also spread increasingly to the maritime and armaments industries, and it played a major role in tools. Metal enabled the development of instruments furnished by mechanics, although production of them was hindered by a shortage of wood, especially in Britain.

Studies in France of the commercial use of iron forges in Normandy and Franche-Comté have revealed the importance of these industries in homes, workshops, and farms. In fifteenth-century Franche-Comté, forges were utilized to manufacture Burgundy's artillery and to equip the salt mines of the Jura Mountains. They also served to produce metals for mills, churches, and castles like the one high in the Jura where in 1486 an architect suggested that a neighboring blacksmith supply door fixtures and tools like crowbars, as well as coins, hammers, and mallets for the extraction of stones and the work of masons.[4] The profession of making heavy clocks of metal would have been impossible without the transformation of iron production. In sixteenth-century Normandy, "the rise of metallurgy was certainly reflected in the increased use of iron and cast-iron products."[5] An entire artisanship of metal production by manufacturers of locks, clocks, and arms was created around French forges.

In the eighteenth century, raw iron from the Franche-Comté was exported to other regions to be refined into steel for scythes, sickles, sword blades, files, and other products. The factories of Saint-Étienne transformed iron from the Franche-Comté into small utensils, while the tilt-hammers of the Dauphiné converted cast iron into large ones and the workshops of Vienne turned out wheel rims, knife edges, and locks. The best cast-irons of the Saône Valley were sent to Toulon to arm warships. The utilization of iron in the construction of buildings spread rapidly in the eighteenth century. Many edifices were reinforced with iron to increase their strength, for example the dome of the church of Sainte-Geneviève in Paris (later to be the Pantheon) constructed by Jacques Soufflot in 1757.

Previously mentioned was the British development, since the beginning of the seventeenth century in the regions of Birmingham, Sheffield, Manchester, and Nottingham, of the production of new objects for decoration of the home or the table. In the eighteenth century this manufacturing considerably raised the quality of life. Meanwhile, iron and cast iron were more frequently used in construction. A genuine architecture of iron was born.

Transformations thus occurring from the seventeenth century to the end of the eighteenth led to the metallurgy of the first industrial revolution, which was marked by the use of the reverberatory furnace, the production of cast iron, the brazing of metal, and the rolling mill. The slowly spreading use of mined coal in various British industries has been described. English workers acquired unequaled know-how in mastering fire, in terms of both the choice of materials and installations, such as melting-pots, and the means to load coal and manage flames. A crucial step was taken in the second half of the seventeenth century with the arrival of coke. Other innovations appeared in glassworks, like the process for flint glass, the use of coal, and the fusion of copper by employing a reverberatory furnace. These methods were subsequently applied to all industrial furnaces and to iron puddling.

In 1708 Abraham Darby, who had worked in a malting plant and a brass works, was the first to melt cast iron using mined coal in a blast furnace. This new process was not really adopted until the mid-1700s because it depended on a smelting process that Darby kept secret. Moreover, improvements necessary for both coke production and the furnace function had to await the utilization in 1725 of John Wilkinson's steam engine to operate bellows.

Also in England in the 1760s, the brothers Charles and John Wood perfected a procedure using a reverberatory furnace to de-sulfurize cast iron and large hammers to beat refined iron, which proved to be a great success. In the 1780s, Henry Cort invented a puddling that could replace the blooming of metal. Obviously this process took a cue from the mashing of minerals during their fusion, used to refine metals. It provided manufacturers of iron and steel with a gamut of products for wider use.

The Nineteenth Century

The expansion of metals consumption that began in the eighteenth century accelerated in the nineteenth century, and new metals made their debut, substituting for other materials or replacing other metals. In the eighteenth century, copper replaced wood or lead in ship hulls. Iron was preferred to wood or stone for numerous uses in construction, public works, artisanal and industrial tools, mechanics, household appliances, and so forth.

In the production of iron, the English process with blast furnaces spread onto the European Continent. Certain countries, like Sweden, and some regions, like Burgundy, long resisted using mined coal, choosing to improve techniques with charcoal and hydrological power. In the case of Sweden, that reluctance lasted to the end of the nineteenth century. This survival of old procedures can be explained by the difficulty of using new methods to produce metals of a quality equal to some iron whose superiority for various uses was recognized worldwide. The foundation of English procedures was not shaken, and output was greatly improved through experimentation in factories and development of a metallurgical science dating back to the eighteenth century with close ties to Lavoisier. The utilization of hot air in blast furnaces, conceived by the Scot James Beaumont Neilson between 1828 and 1830, required great improvement in the heating system as well as recuperation of gas from the furnace-mouth, used to operate machines. Cort's puddler oven lost a quarter of the metal. It was improved by replacing sand with cast iron and adopting a madder dye that modified the nature of the oxidant. A puddler's work was virtually inhuman, but the mechanization attempted after 1857 remained imperfect. Rolling mills could produce only metal sheets whose form was previously fixed, and the use of hammers persisted. François Bourdon's construction in 1841 of a steam-hammer weighing 2.5 tons resolved the problem posed by the lack of hammers for constructing large pieces, such as those in naval vessels. Hydrologic presses appeared at the same time, opening a promising new track.

The use of iron in construction became common. Problems with the dome of the Pantheon were solved by iron triangles forged into masonry, a precursor of reinforced concrete. Better techniques in the manufacturing of iron and cast iron after the eighteenth century enabled them to substitute for wood and stone in many instances. Resistant to compression and traction, these metals reduced the danger of fires, and by reducing the supports needed to build warehouses and factories, they offered increased space and better lighting. Initially in England, then worldwide, architects incorporated iron columns and joists not only in warehouses and industrial buildings but also in churches, theaters, libraries, markets, and railway stations. Iron structures, bridges, and above all viaducts clustered along railway tracks after 1850. Engineers perfected techniques of prefabrication. Iron trellis girders were adopted, as was assembly with rivets, improved by the Frenchman Eugène Flachat. Suspension bridges appeared in France and England in the 1820s. Suspension by cables, adopted in France by Marc Seguin and his brothers for the bridge at Tournon on the Rhône, replaced the older system of chains and became a great success. An accident at Angers on 16 April 1850, in which more than two hundred soldiers died, halted construction of suspension bridges in France for forty years, but they continued to be built elsewhere.

Iron made available to engineers in the 1850s proved ill-adapted to some of the new purposes, such as railway switches, parts for locomotives or textile machines, heavy artillery, and armor plates. This led to a radical change in the relationship between clients (enterprises) and suppliers (ironmasters). Starting in the 1850s the market trend moved progressively from sellers to buyers. Until the 1840s the quality of iron and consequently the nature of its use had for generations been determined by the place of origin or by the process at manufacturing sites. Especially sought after were irons from Berry or from the Franche-Comté. This changed with the advent of railroads. None of the iron manufactured in France, England, or Belgium had the quality needed to withstand the shocks of a moving train. Engineers discovered that whereas rails did not oxidize as they feared, they did fracture and were rapidly destroyed. On the railway line from Lyon to Saint-Étienne, opened in 1832, it was necessary to replace the tracks every three years, and broken axles were frequent. One such breakage caused the first great railroad disaster in France when a local train caught fire at Meudon on 8 May 1842 and fifty-five passengers were burned alive. In March 1847 the Northern Railway Company, begun in April 1846, released a report to its stockholders: "Faulty tracks are so frequent that they pose a constant danger for which our company cannot accept responsibility." The relationship between client and producer was from the outset a bone of contention.[6]

Scientific committees were appointed to devise a response to these malfunctions. Intellectuals devised theories, but science had no answer. Railway companies were obliged to adopt a more empirical approach to the problems: metals from different suppliers were compared; operational modes for manufacturers were specified; engineers were recruited from metal factories to oversee the proper implementation of these regulations; metals underwent rigorous testing. A breakthrough came from the French navy, whose mechanics laboratory, created in 1852, was placed in the service of industry to measure products' conformity to contractual specifications. The railway companies adopted these quality tests in the 1850s by founding specialized services and opening workshops that soon became laboratories. Tests under cold or heat principally concerned traction, but also stamping, shock, flexion, and torsion.

Manufacturers were thereby placed under controls, a measure strongly criticized by engineers such as Charles-Henri-François Couche, who in 1867 affirmed that "it is only in France, the classical nation of regulation, where manufacturing has been set back." According to Couche, engineers controlling railways hindered rather than hastened progress; it would be better to organize a real competition among producers. The Northern Railway Company drafted market procedures so as to retain a certain number of "well established" factories and workshops. It thus created a small group of "experienced suppliers," always available, with whom it could maintain effective technical cooperation. Thereby a process of collective innovation emerged from the technological collaboration of clients and manufacturers.[7]

The production of much cheaper steel on a grand scale was another response to the difficulties encountered in the utilization of iron. The initial process, discovered by Henry Bessemer in the late 1850s, required minerals of very high quality. In 1864, Pierre Martin, scion of a metallurgy dynasty, converted a furnace invented for the glass industry by the German Werner von Siemens in 1862, adapting it to work with metals. Finally, a process invented in 1878 by Sidney Gilchrist Thomas and Percy Carlyle Gilchrist resolved problems arising from the presence of phosphorous in certain minerals, thus allowing exploitation of mines in Lorraine. The use of steel to manufacture railroad tracks and for other purposes expanded rapidly after 1870. State administrations and private railway companies played a major role in this evolution by financially supporting enterprises that adopted the new procedures and intensifying competition among them.

The pressure thus applied to producers had two consequences. First, some metal firms in the 1880s—Schneider at Le Creusot and the Forges d'Imphy, for example—developed genuine research activity to improve the quality of metal products and alloys. Next, a science of metallurgy arose, particularly illustrated by the work of Floris Osmond and André Le Chatelier. A long-term dialogue arose between rail companies and manufacturers, despite the climate of tension in their relations. Between the two world wars, this latent conflict became a cooperative research between two partners. A unified committee was charged with the scientific identification of causes of cracks and fissures and the perfection of production procedures. Thermal treatment of tracks became a major object of research, resulting in the creation of the Institut de Recherche de la Sidérurgie (IRSID). Until the 1970s French metal firms entrusted most of their fundamental research to this body.[8]

Generalizing the Model

The history of European metallurgy supports the thesis Eric von Hippel presented in his work *The Sources of Innovation*. He has shown that the most pertinent variable for explaining successful innovation was a clear perception of consumer needs.[9] In the sector of scientific instruments, he analyzed 121 innovations marketed between 1929 and 1934, of which fourteen were classified as elementary, sixty-three as minor, and forty-four as major. In 81 percent of cases, the consumer dominated the process of innovation. It was the consumer who perceived the need for an advance, invented the instrument, spread the information, and marketed the product. A similar domination existed in the industry of microprocessors. In a study devoted to semi-conductors and integrated circuits, Hippel showed that the firms and individuals who developed a prototype and then marketed it were also consumers 56 percent and 67 percent of the time respectively. Because these consumers themselves had the capacity to identify existing unsatisfactory products and define their inadequacy, they therefore had every chance to find responses to such new needs. According to Hippel, the pre-eminent role of consumers also applied in manufacturing, especially in materials industries. In the production of tractor blades, for instance, the principal source of innovation was the consumer in 94 percent of cases, which also doubtless went for machines used in public works. In thermoplastics, 43 percent of innovations were developed by consumers, 14 percent by producers, and 36 percent by suppliers. In general, one must conclude that comprehending the needs of consumers, taking into account their dissatisfaction with old products, constituted the basis of all innovative policy by permitting them to become directly involved in the research of new solutions.

Producers and Consumers in American Metallurgy

In the United States, the oligopolistic structure of metallurgy, far from favoring the rise of research and innovation, actually slowed its development. By 1930 only fourteen of the thirty-three leading producers of steel had created industrial research laboratories, and they were small in size. Active research in the steel industry was initiated not by producers but consumers, who established their own industrial research laboratories and encouraged cooperation among enterprises to solve the technical problems they had encountered with products: "[T]he interaction between consuming and producing steel firms incited efforts of retroactive research in which they worked together to conduct research and to improve the quality of products."[10] The most important of these consuming firms was the Pennsylvania Railroad, which had created two research laboratories in the 1870s with the intention of developing new products and improving existing ones to assure that "railway operations could be more secure and economical."[11]

In the short term, these efforts had the effect of making testing procedures more rigorous. But the strength of consumers was insufficient to compel manufacturers to actually modify their methods of production. At the turn of the century they were still using the Bessemer process and opposing standardization of products. In 1902 the American Society for Testing Materials (ASTM) was founded and became a member of the International Association for Testing Material (IATM), created in 1898. The IATM aimed to promote the scientific analysis of materials and to satisfy consumer interests. In 1903 the American branch had 431 members, including sixty-nine representing producers, forty-four railway companies, sixty-four universities, and seventy-nine business firms. The number of consuming enterprises subsequently grew rapidly.

The ASTM proposed to develop tests and specifications for the principal materials used by American industry and to define those accepted by producers and consumers. To achieve this program, the ASTM put into place an institutional framework designed to establish effective and permanent relations among the different partners in industrial research. From the beginning, steel was its main preoccupation. Very lively clashes pitted producers against consumers in identifying the causes of damaged railroad tracks, but these debates ended with the expansion of cooperative research. The problems posed by defective rails were still one of the principal challenges of research in the 1920s. The experience of cooperative research was extended by the internal research of producing firms, such as the laboratory located within the U. S. Steel Corporation in the middle of the 1920s.

Alcoa and the Aeronautics Industry

The most utilized materials in the fledgling aeronautical industry were bamboo, canvas, and copper wire. The use of wood spread after 1900, but the supremacy of monoplanes favored the adoption of metal, which was initiated by Hugo Junkers and Louis Charles Breguet during World War I. Aluminum alloys soon competed with steel. The use of metal was advanced by the image of modernity it enjoyed compared to wood, whose faults (lack of homogeneity, short life span, need for trained workers) gradually became apparent. As for the advantages of aluminum, they long remained hypothetical while its manufacture slowly improved. The Breguet 14 aircraft could be entirely constructed of this metal in 1916. Aeronautic construction was entirely of metal by 1930. Aluminum possessed qualities of lightness, longevity, and durability. Its adaptation to the demands of aviation nonetheless required a massive research effort. The emergence and development of research in the first American aluminum enterprise, Alcoa, was the direct consequence of a desire to respond to aviation needs.[12] From its founding in 1888 until World War I, the strategy of the firm's directors remained centered on the production of primary aluminum and competition with other products. This research retained a strictly pragmatic character. During the war the firm had to meet the demands of the navy. It was the relationship with this "demanding client" that raised questions about corrosion and led to creation of a quality control service that imposed increasingly rigorous specifications.[13] Aviation became the major customer for high-resistance alloys, used at first for dirigibles and then water planes. The solution found to prevent corrosion was vitally important for a growing fleet of flying machines, and this problem was the starting point for Alcoa's efforts to perfect metallic properties.

In 1919 Alcoa created its research bureau, whose direction was entrusted to a renowned scholar of electrochemistry, Francis Cowles Frary. It was paired with a technical direction bureau. Frary pursued a strategy of developing new products, and in response to questions posed by aeronautics, he elaborated a fundamental research program concerning the structure of materials. In 1927 Alcoa unveiled its first great innovation in aviation, the steel plate Alclad, which assured protection against corrosion. Other advances followed, for instance an aluminum-copper alloy in 1931 that could be substituted for Duralumin, invented by Alfred Wilm in 1911. This alloy inaugurated a new era in the history of aeronautic construction by making transatlantic flights possible. This success continued during the 1930s and World War II. Above all it helps to explain the development of highly sophisticated research in electronics. A new laboratory founded at New Kensington in 1928 demonstrated the fundamental role that "client relations" played in research strategy. Engineers were trained for full-time contacts with builders, thus creating the "technico-commercial" actors whose role was to assure the coordination of market needs with research. A technical committee was created in 1931 to associate researchers with commercial services.

All in all, the "success of Alcoa in the manufacture of modern high-resistance alloys for aeronautics was due in large measure to efforts of coordination by technico-commercial engineers" and "to planning efforts by the same organization to translate the needs of clients and the technical preoccupations of different functional entities of the firm into realizable and hence dependable research objectives."[14] The history of the research Alcoa pursued between 1930 and 1970 to resolve the problems posed by corrosion under tension illustrates this process perfectly. It shows "how the evolution of client needs influenced the development of alloys and how technico-commercial engineers assured the exchange of indispensable information between Alcoa and the aeronautics industry." But "it also shows the interdependence of elements in a chain of technical functions and of numerous specialists involved in the process of improving technology," whether those specialists belonged to the enterprise or not.[15]

The technology of materials was mobilized for many other purposes. Armatures were covered with the stressed skin of a light alloy reinforced by girders. By 1939 nearly all airplanes, despite their infinite diversity, were constructed according to this principle, with increasingly robust elements whose adjustments were more and more precise. Metals reigned unchallenged until the 1970s, when the qualities of so-called composite materials with a base of plastic, especially lightness, were sufficiently enhanced to justify their use in certain components. Composites made up 9 percent of the Airbus A 310 in 1982 and 19 percent of the A 320 in 1988. Today some people dream of aircraft made entirely of composite plastics, of which several prototypes have already been constructed.

As a whole, the history of materials cannot be understood except by accounting for the use of objects in which they are integrated. Producers cannot correctly perceive the needs of consumers of these objects unless they themselves manufacture and utilize them and enter into a permanent dialogue with consumers. The pressure consumers exert on suppliers to adapt products to their uses has constantly oriented the evolution of metallurgy since the Middle Ages. Accordingly, the success of the German chemical industry in the second half of the nineteenth century can be explained not only by the implementation of a scientifically based system of research but also by a clear perception of industries' expectations and by the application of a marketing policy based on knowledge about the consumption of dyes and pharmaceuticals.

Chapter 9

Strategies and Social Networks

The dialogue between producers and consumers is only one of the relationships an enterprise maintains with other components of society, that is, with the networks that form within or around it, of which it becomes an element. The most visible of these networks are those that develop to assure the emergence of a new technology, whether on a global or local scale. The former scale constitutes a worldwide community; the latter, a social group. Less visible but of vital importance are the networks that appear inside of enterprises, between enterprises, or among enterprises and other economic and social actors in their daily or strategic activities.

Global Communities: Gas, Electricity, Automobiles

When a novelty such as electricity or the automobile emerges within a technological system, its promotion is often undertaken by actors from different horizons: enterprises, universities, administrations, social movements. Relationships are based on common values and interests created well before the actual appearance of these technologies. Commonalities are nevertheless an essential factor in the early development of a technology and its stabilization as a sector of activity extended by regulation.

In the 1830s a community concerned with gas began to form in France. Composed of engineers, entrepreneurs, and investors, it took shape around a group from the École Centrale in the 1850s. The Société Technique du Gaz was created in 1874. One of its principal goals was to facilitate the recruitment of personnel, and in 1892 a professional management syndicate appeared. The two organizations merged in 1927. The *Journal des usines à gaz* was first published in 1877 and became an instrument for promotion and defense of this technology, which after 1870 suffered from merciless competition with the new technologies of oil lamps and electricity. Contributions to the journal offered advice for the utilization of appliances in order to stimulate demand. Faced with competition, the gas community stoutly claimed the superiority of its products and directed its research efforts accordingly.[1]

Electricity was the mutation between 1830 and the 1890s that transformed physical science into a preponderant industrial sector spurred by an ensemble of actors united by their passion and mutual interests. The dominant force in this community was a group of intellectuals and university professors like Antoine César Becquerel, his son Alexandre Edmond Becquerel, and Antoine Breguet, who gathered inventors and talented researchers such as Théodore du Moncel and Zénobe Gramme, manufacturers of scientific instruments like Gustave Froment and Louis François Clément Breguet, professors from the Sorbonne or the École Polytechnique, telegraph executives, industrialists, bankers, and journalists. They met in various places and thus all of them circulated "in the Breguet workshop for phototonic sessions."[2] Other locations served for meetings of these various groups, such as the Académie des Sciences, the Conservatoire National des Arts et Métiers, the office of railway companies, or other enterprises using electricity, such as Christophe. The telegraph service was converted into the electric telegraph in the 1840s. The École Supérieure de Télégraphie was founded in 1878, and the telegraph service was attached to the postal ministry, whose inspectors were the first state corps related to electricity. Popular literature meanwhile assured "a general diffusion of talk about electricity."[3]

In 1881, the first Paris exposition devoted to electricity illustrated the cohesion of this milieu. Its organization was delegated in 1880 to a committee presided over by the minister of the postal and telegraph services, Adolphe Cochery. The committee included politicians, engineers, scientists, financiers like Alphonse de Rothschild, and journalists. The exposition was largely financed by private industry and thus appeared to be "a vast operation of marketing" put on by a group of persons "directly interested in the promotion of electricity."[4] It reflected the diversity but also the coherence of the electricity community and revealed its international dimension. It also contributed to unifying and creating the institutional instruments for its expansion. The electricity community became organized after 1881. In 1883, a group of French industrialists, engineers, and scientists established the Société Internationale des Électriciens (SIE), which undertook the creation of a central laboratory for electricity, using receipts from the Paris exposition. Its principal activity was oriented toward "tests performed by the industrial service such as standardization of electrical apparatuses, measurement of electrostatic capacity, and determining the endurance of electrical appliances." This was the beginning of a movement for the standardization of electrical measuring instruments, which led to the creation in December 1895 of a school that became the École Supérieure d'Electricité.[5]

The promotion of bicycles was the result of close cooperation within a group of industrialists dominated by Édouard Michelin and several journalists, of whom the most notable was Pierre Giffard. In the 1890s, the first steps of the auto industry were guided, in France as in the United States, by associations whose prototype was the Automobile Club of France, founded in 1895 and boasting a thousand members by 1897. One of its founders, Henri Deutsch de la Meurthe, was a major organizer of car parades. The first auto race occurred in 1894, and the first car show opened in 1898. Races were not only a way to advertise an automobile; they were also a means of conducting tests and orienting auto technology. They allowed the presentation of new models as well as the exchange of information. It was in France that automobile journalism supported by constructors made its debut. In 1900, auto builders created *L'Auto*, a journal presenting itself as a defender of driving and cycling that had a circulation of 13,500 by 1913. Besides organizing extensive propaganda campaigns favoring the use of autos, automobile clubs and constructors participated in the creation of models, influenced public officials, gave tips about driving, and published technical journals.

Social Groups

A technological framework highlights the established interconnections among the different actors—scientists, engineers, entrepreneurs—who participate in the emergence and development of a new technology. These various actors form a social group whose goal is to promote a particular technique whose success is in everyone's interest. The degree of these actors' inclusion within a technological framework and a social group is quite variable from one actor to another, and each actor may participate in several groups. According to Wiebe Bijker, "the inclusion of actors in a technological framework may be characterized by their goals, their strategies to resolve a problem, their experimental know-how, their theoretical education, etc.," and "the degree of inclusion of an actor is not constant and may be altered in the course of events."[6]

The History of Bakelite

Bijker uses the history of the invention of Bakelite to illustrate this concept. In this analysis, Leo Hendrik Baekeland was involved in the 1890s with a social group constituted within the framework of celluloid chemistry and electrochemistry. Members of the former group had the objectives of producing plastic objects principally for consumers, modifying unduly dangerous manufacturing procedures, and finding new commercial outlets. At the heart of this group, Baekeland developed an original research strategy oriented toward products destined for industry rather than the consumer. His inclusion within the celluloid community was gradually reduced until 1907, when, after submitting his major patent for Bakelite, he undertook the formation of a new technological framework. The productive procedures Baekeland introduced were so different from those of the technological boosters of celluloid that they could neither understand nor concede the advantages of his work.

To resolve the difficulties encountered in the production of celluloid, Baekeland set a unique course by combining current theories of inorganic chemistry with the study of fluids and electrochemistry, and by conceiving on this basis new and coherent strategies of research. To transform laboratory research into industrial success, he launched a strategy that could be defined as "social construction." His initial social group was limited to employees of the Bakelite Corporation, but other manufacturers gradually joined in, "their methods and concepts being partially integrated into the technological framework of Bakelite."[7] By 1912 the management of the American General Bakelite Company was almost exclusively composed of former competitors. A second phase was the engagement of users of the product, in particular the social groups around automobiles and radios, since Bakelite was an effective electrical insulation. Accordingly, it led to auto components assuring headlighting and easy starting, and it could also be used for non-electrical parts of the vehicle. A third phase was the multiplication of usages for radio as well as for industrial and domestic purposes. Bakelite thus became a universal material. To develop its uses, Baekeland cultivated relations more often with engineers than with managers. He encouraged the emergence of a new technology of Bakelite molders, and he solicited other actors—industrialists, publicists, artists, designers, journalists—who conceived, produced, and promoted new products. He developed a luxury trade and the effective marketing of it. During the 1930s, the Bakelite network stabilized. Its members developed solid connections resting on the successful promotion of a product whose uses seemed unlimited.

The History of Oxy-Acetylene Welding

The collective character of technological systems is well illustrated by the emergence of the technology of oxy-acetylene welding between 1897 and 1930. Two symbolic events marked this accomplishment: the creation of the French Institut de la Soudure, devoted to teaching and research, and of the École Supérieure du Soudage et de ses Applications (ESSA). These two organisms were financed by a cartel, the Comptoir des Producteurs du Carbure Calcium, and by the business enterprise L'Air Liquide with its affiliate, the Soudure Autogène Française.

The opportunities offered by the new welding procedure that appeared in the 1890s concerned these "social groups," to use Bijker's terminology: the producers of calcium carbide, the producers of acetylene, and the producers of oxygen. One of the main stakes in the relations among these professional groups was control of the production and utilization of blowtorches until the early 1900s, when "welding gained its independence."[8] In fact it was brought at this time into an effective professional organization supported by producers of calcium carbide and by a flock of companies controlled by L'Air Liquide, thereby creating the genuine institutionalization of a technological framework of welding and of a social group of welders formed around this great enterprise.

The starting point of this new development was Henry Louis Le Chatelier's work on high temperatures. In 1892 Henri Moissan perfected the production of calcium carbide in an electric oven. Transposed to industry, this procedure allowed inexpensive production of acetylene used for lighting—which, however, remained dangerous. In 1896, Georges Claude and Albert Hesse, engineers at the Thomson-Houston firm, developed a manufacturing procedure for acetylene that considerably reduced the risk of accidents, which Le Chatelier further improved in 1897. Meanwhile, his work permitted the use of acetylene to replace oxyhydrogen or oxygen blowtorches.

Faced with these discoveries, the various actors organized. A brother and a son of Henry Louis Le Chatelier founded the Compagnie Française de l'Acétylène Dissous (CFAD), which supported Georges Claude's research on liquefied air and that of Charles Picard and Georges Fouché to improve the oxy-acetylene blowtorch, perfected in 1901. To finance research that might allow industrial production of liquefied air, in the same year two members of the Thomson-Houston company, Frédéric Gallier and Paul Delorme, created a "financial syndicate" that became a firm called the Société de l'Air Liquide to study the Claude process, not definitively perfected until 1905. This was the beginning of a "fabulous industrial adventure."[9]

Several companies producing acetylene, either affiliates of the CFAD or independent enterprises, were formed in 1898. André Le Chatelier, an engineer with the French navy, succeeded in 1903 in obtaining authorization for the use of acetylene welding for repairing ships "in which the durability of welded joints was not in question."[10] He also created the Société d'Acétylène Dissous du Sud-Est (SADSE) to exploit this procedure, whose deployment was particularly developed abroad after 1906. In response to a concentration of business firms in 1909, André Le Chatelier created the Soudage Autogène Française (SAF), which rapidly became the premier welding company of France, though it was actually under the wing of L'Air Liquide. Paul Delorme, after becoming president of L'Air Liquide, entered the executive council of SAF in 1911.

The SAF bought the patents of other welding procedures, notably that of the Swede Werner Kleeberg's electric arc welding. The executive council of SAF stated in 1920 that these "fusions and purchases have permitted SAF to maintain the supremacy of the French business of oxy-acetylene welding of metals."[11] During the 1920s, SAF thus sustained a dominant position in this expanding market. The utilization of welding technology spread to railways as well as to aeronautics, automobiles, and metallurgy. SAF created a laboratory in Paris and maintained close relations with the scientific community. Thus, in effect, "the scientific bases of oxy-acetylene welding were founded," thanks principally to the work of Albert Portevin.[12]

L'Air Liquide's ambition to dominate the welding sector was furthered by the construction of an institutional network largely controlled by it and devoted to supporting the development of acetylene welding. This was accomplished by superimposing patron associations created to defend and organize the sector, a strategy adopted in the 1890s. A decisive step was the opening of an Office Centrale de l'Acétylène (OCA) in 1905. The OCA was taken in hand by Raphaël Granjon, founder in 1898 of the Union des Acétylénistes du Midi, and Pierre Rosenberg, editor of the *Journale de l'acétylène* and secretary of the Union Française des Acétylénistes, created in 1902. The OCA soon acquired an information and publication service, a documentation center, and a research and testing service. It also oversaw the quality of installations. In 1909, Granjon was designated president of the Union de la Soudure Autogène. Thus, before war broke out in 1914, L'Air Liquide's control of these associations was in place.

The OCA was structured during the war and the 1920s. Granjon instituted national and international organizations, distributed a large number of publications, organized congresses and meetings, started programs of collective research, and above all developed teaching and education. This action climaxed at the end of the 1920s with the foundation of the Institut de la Soudure in 1930, which set as its objective "the study and defense of the economic, industrial, and commercial interests of the various procedures of welding."[13] In February 1931, this institute received recognition from the École Supérieure de Soudure Autogène. It was chiefly through this school that the French sector of welding was able to construct its identity and strength at the national and international levels. This collective success was inseparable from that of the Société de l'Air Liquide. As for calcium carbide, it soon ceased production.

Hence, within thirty years an organized social group formed around two collective institutions—the bodies defending calcium carbide of acetylene and then oxy-acetylene welding—so that a great enterprise became multinational. This result was the culmination of a process of social construction that involved actors with often very diverse interests. There were the brothers André and Henry Le Chatelier and their entourage, the producers of calcium carbide, university professors of various disciplines, naval engineers and shipbuilders, functionaries of the Ministry of Commerce, and the staff of L'Air Liquide. Yet this accomplishment was also the product of obstinate action by certain particularly determined actors who applied a winning strategy with sangfroid and assured the worldwide success of a technology, a professional sector, and an enterprise. Obviously they thereby composed a "social group."

Enterprises and Networks

The concept of a network applies equally well to the organization and functioning of an enterprise and to its activity as a social group. Thus, it concerns the model of an enterprise as well as the relationships it establishes with other enterprises and other actors in the economy. It is therefore appropriate to distinguish between internal and external networks.

Internal Networks

The model of a large integrated business enterprise appeared in the final third of the nineteenth century and became generalized in the course of the twentieth century. It concerned not only the internationalization of research but also the ensemble of business activities. The organization of an enterprise rested on a multidivisional and hierarchical structure. A system of strict controls assigned to several of its branches maintained unity of the entire system. Alfred Chandler has contended that this mode of organization was more effective than market mechanisms when high capitalistic production was tied to a mass market. The integration of operations within the enterprise allowed it in effect to reduce uncertainties and costs.[14]

The vertical model of centralized and multidivisional organization based on a division of tasks among different services, with a hierarchy of procedures, was challenged in the 1970s. At first the superiority of this model, which assured good coordination among various components, was confirmed by analyses concerning both the creation of tacit learning and the construction of formalized knowledge. All scholars agree that the quality of horizontal relations established among different services or divisions of an enterprise is the primary condition governing its success, a conclusion also clearly apparent in the preceding chapter. This can be analyzed in terms of a network that compensates for the detrimental effects of the isolation caused by hierarchical organization. Likewise, an econometric analysis in 1994 by Nathalie Greenan and Dominique Guellec showed that the differentiation of products is better assured by a "decentralized (horizontal)" organization based on knowledge-sharing among workers than by a "hierarchical (vertical)" organization in which an engineering division defines the work to be accomplished by other services.¹⁵ Starting in the 1980s, these observations justified the adoption of a project-based organization. Involving a global approach to innovation, this organization created transversal teams of researchers and engineers of production or marketing responsible for the conception and development of a product from its beginning to its commercialization. Thus, it replaced the vertical integration model that separated different functions and entailed an absence of collaboration.

In the 1980s numerous enterprises went much farther. They adopted a network organization "in order to vitalize each element of [their] internal structure."¹⁶ This organization imitated the networking of their partners, suppliers, sub-contractors, and consultants. Based on a decentralization of tasks and a breakup of basic units into "centers of profit," this type of management institutionalized transversal teams of conception and production. A hierarchical structure was replaced by "a network of business units, often small in size, either directly attuned to the market and acting as autonomous centers of profit, or serving as technical and administrative support for them."¹⁷ The advantages of this decentralization seemed evident: proximity of tasking, incitement to innovation, creation of know-how, greater responsibility and therefore greater engagement by staff members.

One of the first enterprises in France to choose this model was Merlin Gerin at Grenoble. It was subsequently adopted by multi-technological firms such as Asea Brown Boveri (ABB), AT&T, Alcatel, and Thomson. Pierre Veltz cites the example of Thomson-CSF (wireless telegraphy), which in the 1990s "was entirely reorganized into eighty strategic business units (SBU), forty technical business units (TBU), and forty-two common efficiency teams (CET). The CET were installed to assure the transversal feature of the network by animating common evolutions and capitalizing on common objectives such as technical standards, databases, purchase contracts, and the like.¹⁸ This model even expanded into the chemical industry—until then structured around large basic products—which was reorganized into units defined by market functions. It was naturally applied as well to enterprises with worldwide affiliates such as Lafarge, for which it virtually became standard. However, the dispersal of enterprises had its limits. The multiple transversal contacts among units were difficult to manage, the exchanges among them were not spontaneous, their degree of autonomy was not simple to negotiate, and the relationship between these decentralized units and the decision-making center of operations was a constant source of tension. Some sectors therefore tended to reduce autonomy in order to "combine flexibility with coordination."¹⁹ Various methods of management were devised to assure the control of units by the center: strategic plans, definition of general regulations, guidelines, strict financial rules, definitions of a respectable business culture. But the beachheads of initiative were maintained. A new equilibrium was thereby sought, even if controls were not fully restored and such sectors as communications retained a model of dispersal.

To understand the reasons for adopting a model of business networks, alongside the internal factors one must add two major developments in the economic system: the globalization of markets and the rapidity of technological change, "which constantly threatened obsolescence and obliged enterprises to keep themselves regularly informed of new procedures and products."[20] In this perspective, the internal organization of an enterprise was constructed in the image of its external relations. Internal and external networks were interconnected.

External Networks

Hans G. Gemünden and his coauthors have demonstrated that the innovative capacity of an enterprise is better correlated with the intensity of external connections, clients, research organizations, and other enterprises than with factors like sector or size. Two types of relationships between enterprises can be identified. The first concerns the ensemble of productive activities and of commercialization, which can be called functional networks. The other involves strategies to assure the long-term development of the enterprise, termed strategic networks. The great multinational enterprises that began to form in the 1850s soon established decentralized decision-making within the framework of a vertical structure that in the 1980s approached a business model of networks. They were located at the crossroads of internal and external networks.[21]

Functional Networks

Production includes relationships with suppliers, negotiations for co-production, agreements with competitors, and relations with sub-contractors. Marketing requires contacts with clients and merchants. Functional relationships or networks have developed considerably since the 1950s through the appearance of a brilliant style of management based on the externalization of functions not regarded as strategic. This phenomenon has been extended to all sectors and all countries because of the weight of internal expenses, the difficulty of estimating them, and the wish to escape the consequences of uncertain economic fluctuations as well as the risks of research and the unpredictability of markets.

Relationships between clients and suppliers may occur either in long-term transactions built on loyalty or in competition regularly favoring the lowest bidder. Between these two extremes a compromise might be reached that takes into account both competition and the variables of quality and cost. These strategies determine the structure of the supplier sector. Clients strive to avoid the formation of overbearing monopolies that restrict the transaction to a single supplier. This was the consistent strategy of railway companies in the nineteenth century, and it remains so. Agreements for co-production between competing enterprises are made to coordinate manufacturing capabilities and financial resources in order to realize gains from a particular product, or else to develop a new gamut of products. That is frequently the case in highly capitalistic industries like automobiles.

Definition of long-term strategies principally concerns research and management of knowledge, especially patents and standards. Large enterprises formed after 1890 and throughout the twentieth century have based their power on controlling patents distributed to participants in their network. Research thus provides the means "to circumscribe the technological territory of the enterprise and to protect it." Inversely, the exchange of patents constitutes one of the essential elements of accords between enterprises.[22]

Strategic Networks

Various forms of collective research appeared at the end of the nineteenth century. Collaboration among firms and institutions became established over increasingly long periods through contracts and accords often intended to accomplish programs of basic or applied research. Since the 1940s this kind of research has been radically altered, multiplied, and generalized, and the ever more numerous contractual partners have accepted complex agreements with more precise and restrictive procedures.

These changes resulted from transformations in the means of constructing technological and scientific knowledge, and also in the strategies of business enterprises and states. Research has frequently become interdisciplinary, and often these firms' activities have become interdependent. But the duration of a given technological function has become more limited, making it necessary to share the growing costs of research. Competition among business firms, through easing trade restrictions and raising opposition to monopolies, has intensified. And exchanges among actors in real time, made possible by digital communications, have modified research practices by enabling close cooperation among distant and dispersed centers.

Even more radical in its effects has been the role of the state in the rise of collective research, apparent in budget allocations to certain sectors like space, defense, and atomic energy, and by the creation of vast research programs and organizations. Such programs have promoted cooperation between enterprises and laboratories by imposing precise guidelines on them. Moreover, organizations such as CERN in Europe and NASA in the United States have assumed command of a mega-science in their particular sectors. These programs and institutions maintain close relations with enterprises that participate in research, so genuine networks have clustered around them.

In all, collaboration among firms and institutions has been significant in orienting systems of production and breeding innovation, thereby creating new knowledge, new skills, and new activities.

From Multinational Firms to International Networks

The model of vertical, hierarchical, multinational enterprise came into being in the 1850s. The structure of such firms was both centralized and dispersed. Their internationalization was a natural consequence of market growth and a need to control the development of shared technologies. The history of General Electric between 1890 and 1930 illustrates this double strategy. It may be recalled that before 1914, GE, thanks to a coherent central laboratory research policy since 1900, was able to control the world market for electric light bulbs. The firm's share of the American market grew from 25 percent in 1911 to 71 percent in 1914. GE thus obtained the means to conduct a large-scale international research program. It became essential to consolidate the firm's position abroad by spreading the company's research results throughout the world. Some installations, particularly in France, were established as early as the 1890s. The worldwide network took shape with the creation of the International General Electric Company (IGEC) in 1919. GE set three goals: sell its products, protect the American market, and manage electrical construction elsewhere. The organization established was "based not only on the delimitation of markets but also on the technical and financial collaboration of its affiliates and the reinforcement of the weakest."[23] In 1930 GE constructed a network of twenty-three affiliated branches and eighty sales agencies, plus close technical connections with seventeen associated companies like Thomson-Houston in France. These foreign firms pursued largely autonomous research policies, but together they formed an international technological pool never again to be challenged. Thereby combining centralized organization with partially autonomous entities outside the firm, the GE model was located in the seam between vertical integration and organized networks. The opportunity afforded its affiliates to exploit GE's patents constituted the framework of this system.

Worldwide expansion underway as of the 1960s was reflected in a forward leap by international research and in the developing relations among multinational enterprises. These firms created research laboratories abroad and promoted various forms of collaboration between central and local laboratories. A survey of these international networks in the mid-1980s, covering 580 of the largest firms in the world, concluded that five hundred of them had more laboratories in other countries than in the mother nation.[24] A clear difference marked two types of laboratories. In a general but not exclusive fashion, central laboratories concentrated on fundamental research, whereas the scattered local laboratories pursued product development. The former were highly specialized; the latter, mainly diversified. Such versatility, which could be seen as a response to the rapidly changing environment, required a larger range of competence than was necessary for the central laboratory. The foreign laboratories were often integrated into a local network of research and development, and they were more readily engaged in cooperative research than were central laboratories that limited their external contacts to precisely defined programs. Furthermore, local laboratories often maintained close relations with firms' marketing and manufacturing services. Their research strategies therefore appropriated new learning through patents and agreements among enterprises intended to control and retain positions established in various markets. Altogether, networks of technological cooperation were thus created around the great multinational firms on the basis of accords concluded within those networks and among international enterprises.

Techno-Economic Networks

According to Michel Callon, enterprises and laboratories were increasingly integrated into systems of innovation that he has called "*réseaux techno-économiques*" (RTE). These networks, whose structure and composition vary over time, have become "the true actors of innovation." This development followed from two major changes in industrial societies between the 1960s and the 1980s. The appearance of new actors and new research goals meant enterprises and markets were no longer the sole means of coordinating economic activities. It was no more a matter of only constructing codified academic knowledge and certifying or marketing innovations to create competitive advantages, but also of executing actions of general interest by participating in programs for collective research, education, popularization, or consulting.[25] Hence "the unit of reference was no longer the firm, the center of research, [or] the consumer, but a system of coordinated relationships among those various actors." For this reason Callon speaks of "meta-organizations." Meanwhile, "technological objects" had altered in status. More and more active, they played "an essential and irreplaceable role in the economy of these meta-organizations."[26]

Under these conditions, Callon adds, "the notion of networks [was] a good candidate to succeed previous categories such as spheres of activity, institutions, [or] organizations."[27] He shows that the builders of networks had the objective of creating "irreversibilities."[28] From that point of view they were scarcely different from the entrepreneurs and laboratory directors described in the traditional literature. All of them attempted to control the currents of knowledge and technological know-how that other evolving sectors could not neglect.

One should, however, qualify the novelty of these situations. The two transformations described by Callon were not the terminus of developments begun well before the 1980s and 1990s. It would be more exact to speak of tendencies dating back to the nineteenth century, if not the eighteenth. Technological and scientific research has always had an interdisciplinary character, which became strongly accentuated toward the end of the nineteenth century. Technology was ever an essentially political and financial matter, and well-defined technical objects and social projects have always been factors in social and cultural practices, which railways upset with a force much like that of the Internet. Consumers have always used technological objects in unexpected ways. Entrepreneurs and laboratory directors have long taken account of data that were exogenous to market conditions and to their research. The dependency of research and production on changes in the economic and socio-political environment is nothing new. But because this dependency has concerned increasingly numerous sectors of activity and actors, it has become considerably weightier by creating more and denser interconnections.

Moreover, the integration of business firms and public research institutions into social networks, as described by Callon, does not at all signify that their roles have diminished to the point of being only one element among others. Both have managed to simultaneously become actors within networks, as well as sites of convergence and confrontation among exterior networks. Enterprises are different from other actors because they need to retain their employees. The role of business managers in has precisely been to accept their integration into social networks while preserving the autonomy of their enterprise and maintaining the vitality and creativity of accumulated knowledge.

The Pillars of Innovation

The concept of concentrated centers of knowledge advanced by Fernand Braudel has not lost its relevance. In the course of economic history since the sixteenth century, such centers have provided the thrust of growth. Some among them—such as Amsterdam in the seventeenth century, London in the eighteenth, and New York in the nineteenth—were part of a worldwide system of relationships. Others of lesser reach have exercised significant influence in a more limited space. The hierarchical structure developed in those concentrated centers was accompanied by a transportation network that encouraged exchanges of not only goods but also ideas by assuring the circulation of writings, manuscripts, printed books, journals, and correspondence, and by facilitating meetings among scientists, engineers, and entrepreneurs. In the nineteenth century, technological and scientific colloquia flourished at international expositions, congresses, and gatherings organized by professional associations. Nowadays, the new means of information and data exchange provided by telecommunication networks have joined, without replacing, transportation networks.

This has made the creation of innovations and the continuance of high quality possible in particular activities. Some of these centers have been pillars of technology and scientific excellence often associated with strong economic performance. This is surely the case for Glasgow in the 1860s and Silicon Valley today. Other, smaller zones represented local productive systems organized around ceaseless accumulation of know-how by small or medium enterprises that developed their own social practices. The example of Grenoble after 1890 is illustrative of this concept of pillars of excellence.

The Scientific Pillar of Grenoble, 1892–1970

This history may be divided into four episodes.

1890–1914

At the turn of the twentieth century, the industry of the Dauphiné experienced a considerable boom and a reorientation based on the exploitation of hydrological power by local entrepreneurs like August Bouchayer, a specialist in pressure pipes, and André Neyret and Charles Beylier, specialists in water turbines. This industrial awakening was simultaneous with the emergence of more modest yet significant institutions of cooperation between industrial and university circles.

In 1886, Paul Janet, a young graduate of the École Normale Supérieure, was named an assistant professor at the University of Grenoble at the age of twenty-three. In response to an appeal from the Grenoble Chamber of Commerce, he created a regular course in electricity in 1887. Industrialists funded the equipping of Janet's laboratory. In 1898 Janet's successor, Joseph Pionchon, opened an electro-technical institute near the university under the patronage of the Société de Développement de l'Enseignement Technique (SDET).

After 1900, Janet's institute cultivated contacts with industry by integrating entrepreneurs into the teaching corps, such as Georges Routin, chief engineer at the firm of Neyret-Beylier. The institute enrolled 450 students in 1914. In 1907, Casimir Brenier, president of the Chamber of Commerce, made a gift to the institute of two large properties and an abandoned factory, which became a center of operations including a heating plant. This gift exemplified "the careful and consistent character of financial assistance offered by the industrial milieu."[29] Accordingly, the installation at the institute of a station for testing turbines, financed by industrialists, met with great success. Thus, despite the limitations of this collaboration at the time, these relationships laid the basis for more significant cooperation later.

1914–1950

During World War I and the interwar period, contacts were extended and strengthened through research, in which the Institut Polytechnique de Grenoble (IPG) and to a lesser extent the university participated. Hence in 1922 the joint firm of NBPP (Neyret-Beylier and Piccard-Pictet) created the Laboratoire Dauphinois d'Hydraulique (LHD). Other laboratories and institutes ensued. The École des Ingénieurs Hydrauliciens was founded in 1929, followed in 1931 by the Institut d'Électrochemie et d'Électrométallurgie, both of which were incorporated into IPG. Professional engineers lectured at IPG and made their laboratories available to the university.

The birth of this new spirit explains why several talented young academics sought out the University of Grenoble after World War II. In 1945 Louis Néel left the University of Strasbourg and transferred to Grenoble because there a "lively" industrial milieu existed, whereas in Strasbourg it was "dead."[30] As a student at the École Normale Supérieure in Paris, Néel had written his thesis on molecular theory. After the armistice in 1940, he took up residence in Grenoble without a teaching post. Then he created a laboratory of magnetics and in 1941 signed several contracts with industrialists. In 1942 his contract with the electrical firm of Ugine resulted in a patent. Ugine installed a factory in Grenoble to produce magnets. Néel requested a new Laboratoire d'Électrostatique et de Physique du Métal (LEPM), which he obtained in 1946. The next year, he presented a theory of magnetics that won him a Nobel Prize in 1970. During the 1940s, Néel thus brought together different groups into his projects: industrialists, professors, academic administrators, scientists, and the national research council (CNRS).

1950–1970

The Grenoble faculty underwent an unprecedented expansion in the 1950s and 1960s. The number of professors grew from fifteen in 1945 to ninety-three in 1965, and laboratory assistants from twelve to ninety-three. Not alone, Néel was one of the architects of this success. His role as laboratory directory and university administrator permitted him to create "a physics empire in provincial France."[31] His efforts promoted new activities. In 1952 Néel obtained from the École Normale Supérieure the transfer of a research group in magnetics led by Michel Soutif, who opened a laboratory at LEPM. He oriented it toward spectrometric physics and signed contracts with the Société Alsacienne de Constructions Mécaniques (SACM), which became Acatel in 1965, and with the Division Générale de l'Armament (DGA). In 1955 Néel founded the Centre d'Études Nucléaires de Grenoble (CENG), which marked "a major turning-point in Grenoble's growth as a scientific and economic pillar." By 1967 CENG employed 940 researchers, engineers, and technicians devoted to basic research. One of the main consequences of this installation was launching activity in electronics. The electronics laboratory of CENG joined with physicists from a laboratory of solids to create the Laboratoire d'Électronique et de Technologie de l'Information (LETI). A school for electronic engineering was founded in Grenoble in 1957, the same year that an affiliate of the Société Française de Radioélectrique was installed nearby at Saint-Égrève. Thus, CENG was "deeply imbedded in the local, university, and industrial fabric."[32]

In 1956 Jean Kuntzmann founded the Institut de Mathématiques Appliquées de Grenoble (IMAG). Appointed assistant professor at IPG in 1945, he had taught mathematics specialized in high frequency and had worked for the navy, the air force, the Compagnie Nationale du Rhône, Électricité de France, and especially for the firm Neyrpic. IMAG quickly attracted other European enterprises and laboratories.[33] He collaborated with IBM in computer programming in 1961, and in 1968 IBM France installed a new scientific center near Grenoble, which immediately became a computing firm of worldwide importance.

Until the 1960s, according to Dominique Pestre, "the relationship of university and industry was not progressing."[34] Contracts having major scientific ramifications were rare. But Néel's experience was a first step. LETI, for instance, developed research programs at the request of industrialists or devised new products itself before selling the patent. Several other laboratories were also successful as partners with industry or as researchers, particularly in electronics, during the 1970s and 1980s. Similarly, CENG's general metallurgical laboratory passed on its discoveries to industries under permanent contracts signed with enterprises in the nuclear energy sector.

One factor of coherence in the 1960s concerned the activity of the Association pour le Développement de la Recherche (ADR) at the University of Grenoble. Created in 1958, it was long involved in laboratory tests in mechanics and physics at IPG. Half of ARD's board of directors were members of the Association des Producteurs des Alpes Françaises, led by the indomitable Paul-Louis Merlin. In 1963 thirty-five laboratories were represented in it. The total value of their contracts grew from 3.1 million francs in 1956 to 555 million in 1965.[35]

The 1970s

A new era dawned in Grenoble during the 1970s, which saw the installation of two new and powerful research institutions, one created by the American firm Hewlett-Packard (HP) and the other by the Centre National d'Études et Télécommunications (CNET). Meanwhile, the expansion begun in the 1960s accelerated. HP opened its first Grenoble center in 1971, and a second was installed at L'Isle-d'Abeau in 1985. The former was a research site; the other, a manufacturing plant for all of Europe. This enterprise was integrated into Grenoble's university and industrial network. In particular, it absorbed a number of innovative small firms, which is why a substantial reduction of employees in 2003 was regarded as a betrayal. In 1977 the management of CNET created a research site in Grenoble, the Norbert Ségard Center, assigned to the production of integrated circuits in partnership with industrialists. Thus, "an effective network of various actors was put into place for the development of products meeting the needs and interests of CNET."[36]

The creation of small innovative enterprises continued in the following years with the proliferation of cadres in the Mors firm and the Société Européenne de Mini-informatique et Systèmes (SEMS). These businesses seized the opportunity offered by the opening of ZIRST (Zone pour l'Innovation et les Réalisations Scientifiques et Techniques) in the Grenoble suburb of Meylan, an idea born in 1967 within an agency studying urbanization of the region. The directorate of CENG, supported by Paul-Louis Merlin, organized this cooperation between technological and scientific research with administrative personnel. The project rested on two principles: the establishment of a governing committee, and a grant not of money but of open land. ZIRST's first wave of rentals followed in 1972, a huge success that convinced the Chamber of Commerce of the project's soundness. Thus, ZIRST enabled very small firms to start up, "which they would otherwise have been unable to do."[37] The number of enterprises rose from three in 1972 to 138 in 1987, and their employees from 130 to 3,142. Nearly three-quarters of these firms were owned by independent entrepreneurs, the rest by conglomerates. In 1987 30 percent of these enterprises were in computers and 36 percent in electronics and robotics. Altogether the knowledge constructed by the firms ensured Grenoble's successful debut in the realm of micro-computers and highly sophisticated electronics.

In the late 1970s a second wave of enterprises was spurred by a large number of founders from INPG and from big laboratories. Certainly their success was not always assured, but by exploiting new products, some of these new firms were able to find a niche in a particular sector of technology. Today they are still fundamental elements of the Grenoble scene. The city's example shows how different types of knowledge assure the long-term survival of technology that is both intersectorial and interdisciplinary. New needs are created by the birth of new sectors and the development of existing ones. They are also the result of cooperation among enterprises, both local firms and newcomers. Thus the implantation of laboratories and factories from outside plays a considerable part in the renewal of activity. To that end, the locality must maintain its attraction as a pillar of excellence in research and development. The collaboration among enterprises, universities, and public centers of research appeared early in Grenoble, and they have remained essential to its success.

Local Productive Systems

The concept of a *système productif local* (SPL) applies to another concentration of knowledge concerning both know-how and formalized knowledge with a technological component. An SPL may be defined as "a socio-territorial entity characterized by the presence of a community of persons and a population of enterprises in a given geographic space."[38] It is often composed of small enterprises that boast a strong individual identity yet preserve a collective capacity for innovation. The country most clearly marked by this model since the Middle Ages has been Italy, which today has lost none of its creativity. Recently, centers appearing in developing industries have, in their territories, encouraged artisanal tradition, business success, specialized schools, and sub-contracting by large groups. Florence Vidal cites the example of Bologna, which has benefited from a specialized packaging company, developing a chain of food products and then in turn an industry for preparing them. This expansion was extended in the 1920s by the creation of a firm that invented new procedures and machines and trained technicians who founded their own business. This know-how thus produced a complete gamut of widely used innovative machinery.

The success and survival of a region also depend on socio-cultural factors. They are the result of an "atmosphere," to use an expression of the British economist Alfred Marshall, who praised risk-taking and entrepreneurial success as well as original methods drawn from cooperation among different actors, unionization of labor, collective professional action, and social relations.[39] If close coordination among these local actors weakens, the SPL is likely to disappear. If it is maintained, long-term survival is possible, and they may adapt to technological change and the internationalization of markets.

The history of the SPL at Cholet unfolded in two phases. From the nineteenth century up to the 1950s, it depended on two types of labor organizations in the sectors of textiles and shoes, joined by a "common matrix" that mobilized resources "inextricably bound to the territory," by a propensity to create enterprises, and by a uniformity in creating innovations and refusing to lay off workers.[40] In the 1950s the textile sector evolved toward large workshops localized in the town of Cholet and its vicinity, a modernization that was limited by feeble investment and the narrow social context of solely local human resources. By contrast, shoe manufacturing expanded through "mobilization of the territory and coordination of its activities," thereby creating a network of small firms in the surrounding countryside and thus constantly growing.[41] Beginning in the 1970s both of these models weakened. Local textile factories were bought out by enterprises outside the region. This slump was somewhat offset by the rise of the clothing industry, thanks to the initiative of sub-contractors who exploited outside connections "to improve their specialty in terms of jobs, skills, and positioning."[42]

Cooperation and Competition

The conduct of research associated with technological change is comparable in various branches of industry. Within an enterprise and its laboratory, management is organized around a constant exchange of information and various forms of cooperation or competition among actors. The two previously examined sectors, metallurgy and electricity, serve here as illustrations.

Metallurgy is an ancient industry capable of evolving with the changing needs of its clientele. It was confronted in the nineteenth century by a need to adapt its products to increasingly diverse uses and to raise the productivity of workers and installations to meet growing competition. French metallurgists faced demanding customers and extremely severe quality controls. Their response after 1850 was both technological and organizational. They moved progressively from iron to steel, developed closer cooperation with universities, and extended professional contacts among enterprises. This effort led to the appearance of new products that were well adapted to mechanical, industrial, railway, and military uses. Every product had to respond to increasingly precise specifications as industry was gradually subjected to market demands. In large factories, mass production procedures such as steam-hammers and rolling mills evolved from mechanics to electro-mechanics, then on to electronics and computers.

At its beginning, the electrical industry was composed of a multitude of small enterprises closely tied to producers of scientific instruments who actively participated in early innovations. Their cooperation with university intellectuals allowed them to set norms for the development of electro-mechanics and of transportation networks. Starting in the 1880s the multiplication of uses ignited an acceleration of technological competition and concentration, resulting in the formation of large networks of electrical companies. International organizations took in hand the management of standards and of the sector's general problems.

These two sectors' trajectories were dominated by competition among enterprises and laboratories whose development was supported by social groups drawn from all directions. These groups functioned to support the rise of a particular sector that was often controlled by an enterprise. Cooperative strategies were set into place among the various actors, leading to the creation, formally or informally, of interactive networks that aimed to control a territory. Several examples of this process have been sketched. Their strategies assumed that their activities would be local, allowing establishment of the solidarity that was essential to both the enterprise and the surrounding zone. Grenoble provides a remarkable illustration, although its history also shows the difficulties encountered in maintaining the level of excellence necessary for such centers to adapt to international competition.

Part Four

Technological objects are neither socially nor culturally neutral. They are laden with meaning from their conception to their realization. The role of consumption as a motor of technological change will serve to illustrate this fact. In this perspective, such change is the product of individual efforts and dominant collective notions that lead to new cultural practices. The ultimate consumer and the ordinary citizen thus become actors who are no longer passive but active in technological and social transformations. Innovation of a final product or process is thereby an engine of growth.

The two last centuries of European history provide an ideal vantage from which to observe social and cultural behavior in consumption. From early modern times to the twenty-first century, new products were usually intended to satisfy the needs and desires of the upper classes. Their use thereafter spread throughout society in a slow process that made the products, or imitations of them, available in a large number of less expensive versions. Hence consumption underwent constant renewal, and its extension to the lower classes resulted in mass consumption and mass production. This process was enabled by sustained, albeit irregular, growth in proceeds and production, although inequalities persisted. The diversity of consumer behavior did not disappear, despite the existence of a shared model. These transformations were supported by three easily identifiable major aspirations: a general search for the well-being of body and spirit, the availability of communications assuring more rapid and massive circulation of messages, and the exchange of information from person to person.

Four successive models of consumption appeared. Passage from one to another of them resulted from the emergence of new products and new services conceived in response to the malfunctions of the previous model and to the social aspirations that had thereby been created.

1. In the course of the first industrial revolution there emerged, especially in Great Britain, a way of life that sought well-being through the use of new products.

2. From 1830 to 1880 this trend toward a consumer society continued.
3. Starting in the 1880s a second industrial revolution fostered mass production and consumption based on the development of large-scale technological networks.
4. The years from the 1960s to the beginning of the twenty-first century were marked by innovation and triumph in the technology of computers and communications.

Numerous examples will be invoked in the following pages to describe the evolution of these different phases.

Chapter 10

From Early Modern Times to the 1880s

The English Model

The dependence of new products on social needs was manifested with particular intensity in England between 1650 and 1830. These changes resulted from demands for invention as well as creative solutions.

Seventeenth-Century London

Use of semi-durable consumer goods developed among Great Britain's middle and lower classes from the beginning of the eighteenth century. This advance was the culmination of a slow but radical transformation of consumption patterns in part of British society during the seventeenth century. The city of London was at the heart of this change. The English capital was known for its bursting prosperity in the second half of the century, thanks on the one hand to an intensification of foreign and domestic trade and on the other hand to industrial development based on naval construction and the production of consumer goods. London was the first British industrial center to create in its midst a milieu of merchants, salesmen, industrialists, and members of the liberal professions, plus a lower middle class of clerks. The lower classes prospered as a labor shortage spurred immigration and wage increases. Currency circulation increased significantly.

London's prosperity gave birth to a genuine consumer society. This transformation notably improved the standard of living through a proliferation of new domestic objects for the upper classes and production of imitations of those luxury items. Thus, the products that around 1660 had been reserved for the very rich became more ordinary, and a new domestic culture became widespread. This was the case in the development of more refined goods, the use of glass in windows, wallpaper, drapes, fireplaces, and so on. Virtually unlimited textiles became available for clothing, and the use of buckles, brooches, ornamental buttons, other accessories, and stockings spread throughout society, along with cottons and calicos. Artisans and industrialists took the initiative to expand and diversify their markets by creating imitations and thereby lowering production costs, thanks to the use of rather simple machines and the specialization of labor.

Early Modern England

The list of consumer products in daily life continued to lengthen during the eighteenth century, especially in Great Britain. Luxury products such as soap, porcelains, metal objects, and light cloths from the Near East or China were imitated, as were French furniture and glass. Initially designed for the aristocracy and upper bourgeoisie, these items were gradually adapted to the tastes and means of a much larger clientele. Finery could be changed into cheap clothing. The rise in wages created a consumer model based on the improvement of domestic comfort and new durables.

English enterprises excelled in conceiving products destined for both domestic and foreign markets, including culinary items like cutlery, metal utensils, crockery, and ornaments. The city of Birmingham specialized early on in these types of furnishings and especially in copper or zinc buttons covered with an alloy imitating gold or polished steel. Some of these products already represented an initial form of standardization. Clothing was modified by the end of the seventeenth century. Cotton shirts replaced linen, and new twill fabrics were developed for blankets and sheets. The wide distribution of games and toys well indicated the social and cultural aspect of this rise in consumption.

The Industrial Revolution

The term "industrial revolution" designates a major rupture in the rhythm of growth from the 1780s to the 1830s. In fifty years England became more populous and much more prosperous. Whereas previously the annual increase of gross national product had not exceeded 0.6 percent, in this period the English population increased at a rate of 2.7 percent and the annual per capita revenue by 2.5 percent. This leap in national growth was long attributed to a sudden diversion of national income into savings and to investments favoring radical innovation. In reality, the rising rate of investment was quite gradual until the 1840s, when new technologies were created by modern techniques.

From this viewpoint, the industrial revolution appears to have been the result of a slow modification of consumer habits. After the 1760s the pattern of consumption was accelerated and extended to all classes of society. This broader diffusion concerned a range of luxury items in metal, ceramics, and glass, but also watches, clocks, and home furnishings. New household objects that improved comfort and wardrobes spread into the most humble dwellings. Women became consumers, and a culture of consumption centered on the marketplace thus included decorations, domestic items, clothing, and furniture. This outcome was due to new procedures like lamination, stamping, and finishing metals. Maxine Berg has collected 1,700 patents obtained between 1627 and 1825, mostly during the eighteenth century. Sixty-five percent of them concerned metal objects such as locks, buttons, buckles, pitchforks, scissors, nails, fire screens, and stoves. New materials were used: cut glass; leaded glass; sheets of tin; bronze or silver; papier-mâché; gold; steel.

In London, luxury products were manufactured principally by artisans in watchmaking, jewelry, and fine furniture, or by traders of colonial goods. Otherwise, each of the centers of British artisanal industry specialized in some particular type of production. At Wolverhampton locks were made, and in Sheffield cutlery. Leaded and cut glass was produced in Stourbridge, and an gamut of household items in Birmingham. Cheap iron, which reduced the cost of manufacturing finished, industrial, or artisanal products, was decisive in spurring this production. A strong interdependence thereby united the artisanal and modern manufacturing sectors, while the crucial textile industry depended as well on procedural innovations.

The Rise of Mass Civilization in Paris, 1830–1880

In the nineteenth century, the application of the technology of the first industrial revolution created new needs associated with urban growth and with industrialization. The search for comfort and cleanliness, the need for lighting, and the desire to travel were manifestations of this widespread aspiration to health and happiness.

New Homes, New Products

Throughout the century, the appearance of new products fueled mounting aspirations to enjoy more welcoming homes that boasted better light and heat, stricter hygiene, a less filthy environment, and a less hectic life. These wishes undergirded a consumption centered on domestic comfort and urban planning. Population growth plus rising living standards aggravated problems of crowding, especially in a densely populated city like Paris. Social ills reached their peak with the cholera epidemic of 1832. Paris became a city of dust, germs, and insecurity. Urban planning was therefore the municipal authorities' main preoccupation.

Between 1830 and 1880 the creation, diffusion, and imitation of new products and services were sustained by collective aspirations forcefully expressed in an abundant literature written by engineers, intellectuals, sociologists, physicians, and hygienists who militated for genuine social regeneration in order to eradicate misery.[1] Liberal economists in Great Britain and France like Jean-Baptiste Say developed a vision of the possibility of improving the human condition through technological progress and international commerce. The creation and diffusion of new products throughout society was regarded as a source of growth, even if the new model of consumption became distorted by its popularization.

The new fashion of housing that appeared in the eighteenth century was fully realized in the nineteenth. Walls were better decorated and equipment better maintained. Window glass became common. Apartments were larger and colors more prominent. Dwellings contained more decorative and useful objects. Windows were covered with drapes, walls with tinted paper, tables with knick-knacks and statuettes, floors with rugs. These changes satisfied a desire for both comfort and display.

New objects appeared for bathrooms, kitchens, and drawing rooms. Inventors offered a variety of domestic appliances such as the vacuum cleaner and the sewing machine. The latter was the first product for domestic consumption to enjoy broad international success. Bicycles, originally for pleasure, became a means of transportation for all social classes and a new sport, at first aristocratic, then a component of popular culture. In the textile industry, the production of quality cloths continued to thrive. Silk manufacturing in Lyon, for example, was characterized by "a search for perfection by the manufacturers themselves, who specialized and who carefully inspected the quality of primary materials and sought to engage skilled designers and craftsmen."[2] Yet at the same time they succeeded in satisfying the needs of less affluent clients. Thus, under the Second Empire Claude-Joseph Bonnet sold dresses suiting the taste of a clientele who read the fashion journals and wished "to be elegant even though they [were] not wealthy."[3]

In the first two-thirds of the century, cottons were the product most sought as demand rose due to wage increases. In *Le Peuple* Jules Michelet described the seeming revolution in 1842 when the price of cotton dropped. "Millions of customers, poor people who had never purchased before, began a movement ... clothing for the body, the bed, the table, windows; entire classes, deprived since the beginning of time, had them."[4] The jeweler Charles Christofle introduced the technology of electroplating to create a gamut of products that were inexpensive but of good quality. In bourgeois households a mirror placed above the fireplace became indispensable for living rooms. One can only support the verdict of the editor of the *Dictionnaire des arts et manufactures* in 1881 that "the love of inexpensive luxury following the revolution of 1848 was more favorable than harmful for fledgling industry, first by spreading it, then by making it necessary."[5]

The Mechanization of Production

In the eighteenth century, precision mechanics in clock-making, scientific instruments, mechanical toys, and calculators had opened the way general mechanics would follow in the nineteenth century. The steam engine, that primary force of industry, operated machinery and machine tools with gears, cables, and driving belts.

New mechanical devices had to be proportioned with precision. The machine-tool industry responded to this need at the end of the eighteenth century. The boring machine of John Wilkinson in 1775 and the screw-cutting lathe of Henry Maudslay in 1797 were the prototypes. An autonomous machine-tool industry was created in the early nineteenth century and soon flourished in the United States. The speed of machines increased, and they performed increasingly complex and specialized functions. They served to produce machinery of all sorts, like bicycles and then automobiles.

In the sector of cottons, as noted, the process of mechanization began in 1768 with Richard Arkwright's machine that eliminated manual operations in spinning and produced strong threads. Subsequent innovations after 1800 were associated with use of the steam engine. Not until the appearance in the 1820s and 1830s of metal trades were looms widely adopted. In the linen industry, English machinery was highly complex and costly, whereas American machines were quite simple and their products distributed to a mass market. In the spinning of silk, no real change occurred until after 1830, when wooden spindles were replaced by metal ones, which increased their speed five- to tenfold. Silk weaving required an apprentice to operate the warp and weft until Joseph Marie Jacquard achieved the mechanization of this operation in 1801.

The spread of mechanization rapidly took a more radical turn in the United States, especially because of the higher cost of labor. There, machine tools of the second generation were universal grinding machines, milling machines, and turret lathes. The latter better enabled several cutting functions without having to remove pieces of cloth and replace them on the lathe. These machines were adopted in all sectors of the economy. Two other processes had a role in the American system of manufacturing: one was the assembly line, adopted after 1850 by sectors like sugar and paper; the other was the division of labor. In the years to follow, the transfer of these processes to sectors using metal, such as the auto industry, resulted in reducing tasks to a series of always identical gestures and the utilization of very precise machine tools. Thereafter this American model of manufacturing would be transferred to Europe.

Domestic Commodities and Urban Renewal Planning

In the apartments of mid-century Europe, darkness still limited the space of conviviality and the time for leisure and reading. Heating and lighting of public places were therefore objects of scientific research. Darkness in urban areas favored crime and prostitution. In factories, warehouses, and train stations it encouraged laziness and inefficiency. Lighting was the only way to combat the dark, delinquency, and theft. Yet the city was also a place of leisure and luxury. Amusements, commercial activities, and festivals increased the need for light. These general aspirations were articulated in municipal councils and in the press. Hygienists treated light as a top priority.

Fireplaces appeared throughout Europe in the twelfth century, but at the beginning of the nineteenth century they still did not assure sufficient protection against severe cold, and they released clouds of smoke. Since the seventeenth century research had been conducted to improve their function, as summarized in a treatise by the Count of Rumford, Benjamin Thompson (who married Lavoisier's widow in 1805), who concluded the fireplace should have a trapezoid form with exact dimensions and a flue at its base. In the nineteenth century most fireplaces were "rumfordized." In 1826, Nicholas Lhomond invented a sliding shutter of sheet iron that allowed better regulation of the draft and facilitated the use of coal. Flues were also installed, but that technology had reached its limits by the middle of the century. Research turned toward perfecting stoves. Utilization of metal after the end of the eighteenth century assured slower combustion, improved performance, and permitted the use of coal. Efficiency grew in the 1840s with the spread of cast-iron constructions, and metal stoves gradually replaced tile ovens.

Central heating with stoves was nonetheless clearly inadequate. The development of hot-water heating was hindered by a prohibition on high pressure. Steam heating, invented by Sir William Cook in 1785 and highly praised by Thomas Tredgold in 1825, was tested at the Paris Bourse and became a valid science. Hot air, first used in 1806, heated private dwellings, meeting rooms, and factories. In small spaces stationary or mobile hearths were equipped with a covering made of two metal plaques. The solution that prevailed at the century's end was American central heating with low pressure.

Gas was initially used mainly for lighting. Before its adoption, reactions to the fear of darkness were ineffectual despite improvement of candles and oil lamps. Gay-Lussac and Jean-Baptiste Dumas helped to promote the stearic acid wax candle. The oil lamp was much advanced by a timing device, invented in 1800 by Bertrand Guillaume Carcel, that assured a steady flow of oil to the wick. As for oil street lamps, first installed in Paris in the eighteenth century, they numbered 3,600 in 1769 and 12,600 by 1821. Vegetable oil was messy, however, and gave off only a reddish flame. Moreover, it released a foul odor and went out after eight hours. The introduction of more volatile mineral oils simplified the matter and improved the quality of light.

Gas lighting appeared in France and England after 1810. In the 1820s effective methods of purification were perfected. In factories, tollhouses, warehouses, and train stations, gas became an enemy of thieves and chased insecurity and immorality from the streets. After 1840, thanks to gas lines, it entered apartments and public buildings. By 1860 the number of lampposts in Paris had risn to 17,538, and in 1882 to 54,661. In 1855 six existing gas companies were consolidated by the Compagnie Parisienne du Gaz, whose subscribers went from 35,000 in 1857 to 170,000 in 1882. By 1860 industrialists and merchants consumed two-thirds of gas production. Its use was diversified into heating, cooking, and motor power. The Paris gas company purchased the first patent of Étienne Lenoir in 1863 and began manufacturing its own gas motors, which replaced steam engines in small workshops. After 1887 the Otto motor (named for its German inventor, Nikolaus Otto) touched off a huge increase in sales. The electric motor did not supersede gas until after 1900.

The installation of gas conduits was one of the first examples of a specific technological network. Factories producing gas mounted a series of operations that led to a succession of processes such as distillation, purification by condensation or chemicals, controls on production and consumption via gas meters, and regulation of emissions and pressures. It was necessary to determine the diameter of pipes, choose proper materials, and assure regulation of gas pressure in the network. The Paris gas company called on scientists to find solutions for these problems. In Geneva in 1847 Jean-Daniel Colladon and Théophile Pelouze invented an effective procedure for condensation. Alexandre Arson conducted experiments on the flow of gas in pipelines. And in 1860 Dumas and Henri Victor Regnault designed an apparatus to measure the lighting power of gas. Producing such equipment mobilized the best artisans of Paris. In subterranean passages cast-iron pipes were replaced by steel, and lead was used to extend branch lines.

Gas gave off a diffuse light, mixing yellow and violet, which altered colors. Moreover, it was dirty and smelly, and it entailed complex procedures, not to mention the serious risks of fire. It was nonetheless used until the early 1900s and beyond, thanks to the adoption of incandescent lamps in the 1880s. All in all, the conception and regulation of gas networks depended on mastering very diverse technologies under constant change in order to respond to frequent malfunctions and public demands.

Water in Nineteenth-Century Paris

The collection and distribution of water as well as the disposal of used water had to satisfy a pressing popular demand in the nineteenth century. The inheritance from centuries past was meager. The two aqueducts of Pré-Saint-Gervais and Belleville dated from the thirteenth century; the water pump of Samaritaine near the Pont Neuf had been constructed by Henri IV in 1608 and the viaduct of Arcueil by Catherine de Médicis between 1613 and 1623. Three new pumping stations had been installed in 1669 at the Pont Notre-Dame. Finally, at the end of the eighteenth century, Paris adopted the London solution of steam pumps. Operations of the two pumps situated at the Chaillot and Gros-Caillou, initially granted to a water company founded by the brothers Jacques-Constantin and Auguste-Charles Périer, were recovered by the state in 1788.

For collection of water, three propositions competed: directly gathering water from nearby rivers, pumping water with hydrological machines or steam engines, or deploying Roman methods for conveying water with aqueducts from more or less distant sources. In 1803 a Napoleonic order stipulated that all water theretofore shared between Paris and the French state should be controlled by the city. A decree of 19 May 1802 had commanded the opening of a canal diverting water from the Ourcq River to the basin at La Valette. Bonaparte foresaw a canal assuring both navigation and the collection of drinking water. The basin was inaugurated in 1806, and the Ourcq Canal, turned over to a private company in 1818, was finished in 1826. The Saint-Martin Canal would be completed in the 1860s.

In the 1850s Baron Georges Haussmann, prefect of the Seine, adopted the first project to collect water from sources in the Champagne region, first from the Dhuis River in 1865 and then from the Vanne in 1875. A program initiated by state water service director Jean Charles Alphand in 1878 and completed in 1893 increased flows from those two sources and added those from the Arve and the Vigne. Pumping stations were constructed in 1858 at the Pont d'Austerlitz and at Bercy, then in 1883 at Ivry-sur-Seine. They were programmed in 1886 for Ivry, Javel, and Bercy. Thus, Paris had ten pumping stations by 1889 and at the turn of the century a capacity of 273,000 cubic meters of spring water to attain a total, counting river water, of 738,000 cubic meters, as opposed to 150,000 in 1854.

In 1838 only 0.10 percent of the Paris population had running water at home. Most of the 25,000 wells existing in 1831 were out of service or would soon become so because of pollution by cesspools. By 1850 one dwelling in five had a system of water distributed to the ground floor. Water carriers earned their living by drawing water with pumps directly from the Seine or from fountains filled with river water. To provide hot or cold water they circulated with bathtubs.

Most water from the Ourcq was reserved in 1812 for public fountains. Starting with the July Monarchy, the needs of private consumption were better taken into account. From 1800 to 1850 the volume of water distributed from the Ourcq rose from 8,000 to 86,737 cubic meters, of which 30,737 went to private dwellings. Between 1832 and 1850 the number of street fountains increased from 217 to 1,837 and the total kilometers of water pipelines from 39 to 358. The great majority of Parisians received free water. But in 1860 the municipal council signed a contract for distribution with the Compagnie Générale des Eaux, created in 1853. The number of subscribers rose from 921 to 42,520 in 1876 and to 67,800 by 1889. Starting in 1876 the company installed meters. In all, the 69,000 cubic meters of water distributed to Parisians in 1854 jumped to 407,000 in 1885, of which a quarter went to households. By then water carriers were beginning to disappear. Most of the water still came from the Seine, other rivers, and the Ourcq Canal. Spring water was reserved exclusively for drinking. Surface water passed to public services, with the cleaning of sewers and streets given priority. Water purification began, having been studied since the 1860s in the municipal chemical analysis laboratory at the Park Montsouris, whose task was permanent control of water quality from the Seine.

Large nineteenth-century European cities were malodorous, especially Paris. The stench created fears of epidemic, hence the importance attached to sewers. Their length in Paris rose from 26 kilometers in 1833 to 143 in 1853, 773 in 1876, and 1,650 in 1900. They contained used domestic water, rainwater, and street water. Open to visitors, they were regularly cleansed and tested. Draining them was done separately. Primitive filtering septic tanks called *tinettes*, used after 1843, nonetheless allowed the flow of a rather small portion of drainage water. Until the 1850s, the sewers of Paris constituted a single system that emptied into the Seine at the Pont de l'Alma. Later another system was constructed that flowed into the river at Clichy and Assières, which polluted the water distributed to the population. In 1880 400,000 cubic meters of sewer water were dumped into the Seine.

An imperial decree in 1809 ordered the construction of cesspools for water waste. By 1880 they were installed in 70,000 or 80,000 dwellings. Placed beneath basements, they were cleaned every two or three years to rid them of foul odors. Mobile drainage units were authorized in 1834, but, badly maintained, they too became a source of stench and filth. The high cost of *tinettes*, which separated fecal material from sewer water, limited their use to better neighborhoods. Most of the waste was therefore collected by cleaning crews, which in the 1870s removed more than 900,000 cubic meters a year. This material was transported, stored, dried, and converted into fertilizer in refuse dumps that surrounded the capital.

City engineers sought to resolve the problems of rivers polluted by deposits from sewers and of drainage. Progress in microbiology confirmed that diseases like cholera and typhus were waterborne, and that outbreaks were exacerbated by pollution from rivers and canals. Hence water purification became the major objective of research by engineers. They chose a solution of irrigation through permeable soil or sewage farms rather than chemical treatment. A large plot was first opened at Gennevilliers in 1869.

In 1880 the municipal council adopted a project aimed at closing all the cesspools of Paris. That same year an unbearable stench invaded the capital city and revived the memory of great epidemics. Pierre Emmanuel Tirard, the minister of agriculture at the time, appointed a Commission des Odeurs de Paris, of which Louis Pasteur was a member. This commission rejected a project tendered by the mayoralty and favored a filtration system invented by the engineer Jean-Baptiste Berlier, which was approved by the executive council of Ponts et Chaussées. Sewer cleaners unleashed an intense negative press campaign. The Prefect of the Seine convoked a technical committee that condemned basement cesspools, mobile units, and the existing filtration system. In April 1884 the committee decided to authorize "the generalization of complete drainage into the sewer of streets having properly equipped and irrigated galleries."[6] A prefectorial decision of November 1886 declared this procedure to be optimal, but ten years passed before a law made it obligatory and created a tax to finance it. By 1913 24,000 of 71,000 buildings were still not in compliance. In Paris there remained 6,959 mobile disposal units, 25,821 cesspools, and 19,412 *tinettes*.

Despite the hostility of eminent intellectuals and sewer cleaners, Parisian engineers succeeded in imposing a complete sewage system and creating a unified water network through canalizations, watercourses, sewers, and rivers. They thereby assured a cleaner city and better sanitation for its inhabitants.

Mobility: The Railway Revolution

The establishment of railroads between the 1830s and the 1880s unfettered economic and state power. In 1832 a railway line for mining opened between Lyon and Saint-Étienne. A passenger line from Paris to La Pecq, with a road connection to Saint-Germain-en-Laye, followed in 1837. And a law passed in 1842 defined the legal and financial statute and future pattern of a French network. The railway project of the 1840s extended the highway program of Ponts et Chaussées engineers, which had brought spectacular results, establishing a dense network of well-maintained roads to assure safe and comfortable traffic.

Railways were a response to both the difficulties encountered on some routes and the aspiration for greater mobility in all classes of society. Their construction was based on a visionary scheme derived from the Enlightenment and Saint-Simonianism that intended to overcome England's economic lead, render France pivotal in international trade, complete national unity, favor centralized state power, assure universal peace, and promote closer ties among mankind. Everyone realized that railways would be both a source of prosperity and an imposing instrument of power. This program, voted in by the French parliament in 1838, foresaw the construction of a star-like network of lines converging at Paris from the borders and large provincial cities. This centralizing tendency was reinforced by the law of 11 June 1842.

The 1838 law assumed construction and exploitation of the network by the state—which, however, did not possess the needed financial means. Hence the 1842 law announced that the state would fund the infrastructure and private companies would provide the superstructure. Nevertheless, numerous contracts for main rail lines in the 1840s granted the private companies control of all construction and operation, a combination imposing an excessive financial burden on the companies. Their difficulties were one of the causes of a stock market crisis in 1847. After a period of uncertainty, the imperial government of Napoleon III began in 1852 to regroup French railways into six large networks with concessions of ninety-nine years. Thus, in reality each company gained a monopoly in the territory it served.

French industry was ill prepared to fully satisfy the needs of railroads. The performance of steam locomotives rapidly improved, to be sure, but a large gap opened between engines and other technological elements of the rail system. Tracks, traffic regulations, signals, brakes, designs of rolling stock, schedules, and repairs were all technologically deficient. They employed defective materials, unreliable machinery, primitive means of communication, and dangerous procedures. Between 1840 and 1870, however, engineers were able to construct a coherent technological railway system and administration, although both remained fragile. Meeting the challenges required the companies to call on manufacturers to implement important innovations in the 1860s and 1870s, such as new methods of producing steel, substituting mechanical for manual signals, using electrical communications, and employing metal in construction works. Such innovations spread throughout the entire technological system.

Special reduced rates accorded for large quantities of merchandise definitively opened new agricultural and industrial outlets. Railroads thereby brought prosperity, though this entailed the gradual decline or disappearance of many traditional regional activities. The companies paid particular attention to express trains on the main routes, tourist groups, congresses, pilgrimages, and festivals. By 1870 the better trains from Paris reached Lyon in less than nine hours. The conventional wisdom was that it was necessary "to make trips in second class inconvenient and disagreeable, and to exclude any comfort in third-class carriages," although in fact the bourgeoisie used third class despite the discomfort.[7]

The reception of railways in the 1830s and 1840s was not always enthusiastic. A train ride was often experienced as a stressful and dangerous adventure.

Railway pathology even became a medical discipline. In train compartments one might be murdered or have rude encounters: they were ideal sites for sexual aggression, theft, and bad manners. Yet in reality the railroad was a plebiscite: in 1869 more than 110 million passengers boarded trains in France, and there was no doubt about their utility. Not only was the contraction of space unanimously approved, trains afforded easy reunions of scattered families, visits to distant markets and festivals, vacations, and pilgrimages. The installation of a train station on the edge of a town encouraged construction of new neighborhoods and provision of access to them. The station was also a source of architectural innovation, of which the Gare de l'Est in Paris, opened in 1851, was the symbol. Socially, the railway terminal was a frontier between two worlds: one full of sounds, shadows, and lights; the other a place for musing, solitude, departure, reunion, or separation.

Chapter 11

Technological Networks and Communications

The new technologies of the years from 1890 to 1914—electricity, the internal combustion engine, the chemistry of synthetics, biochemistry, radiotelegraphy, aviation, and others—combined to form the civilization of the twentieth century, characterized by a system of energy drawn from fossil fuels, the massive use of electricity and motors, and a technology of materials dominated by synthetics. This system was composed of three sub-systems: large-scale technological networks, mass production and consumption, and complex techniques of information.

French Railways: Rationalization and Cybernetics

A new vision of railway service made its appearance between 1880 and 1914. As one report declared in 1898: "The transportation industry is devoting itself to customers and seeking to facilitate their travels."[1] A price reduction of 40 percent for passengers and freight, made effective in 1892, has been called "a new foundation of French railways."[2] An innovative publicity campaign and new timetables were arranged. Railway coaches became more comfortable, and some trains included corridors with access to toilets. Third-class passengers in 1907 traveled much more rapidly than those in first class in 1887. In 1890 an engineer with the Northern Railway Company, Gaston Du Bousquet, depicted the adoption of compound locomotives as justified by the "public demand" for convenience and comfort. In 1912 another engineer, Henri Bouchard, hoped "that the passenger will feel at ease in trains ... even at the highest speeds."[3] The administrative model in effect before 1914 anticipated a prudent increase in train speed and a slow evolution in the equipment of coaches. For some lines after 1870 the rail companies adopted the so-called block system of spacing trains by distance rather than time. But in the 1890s they rejected an automatic system developed in the United States because of "antipathy for any instrument directly operated from high-speed trains."[4] These reservations about automation disappeared after 1920. Three principles were advanced: rationalization of operations, electrification, and the adoption of automation. After World War II it would be the turn of cybernetics.

Rationalization depended on a new conception of the decision-making process. Planning was introduced in the 1930s, then featured in the quadrennial projections after the nationalization of French railway companies in 1937. A dispatching system and a reform of marshalling yards tested during the war were adopted, as was a commercial strategy based on better understanding of passenger tastes, freight shippers' needs, and net costs. Electrification, restricted before World War I except on lines of the Southern Railway Company, expanded in the 1920s and 1930s. Growing competition from automobiles and airplanes obliged rail companies to make speed a priority. On electrified networks, speed rose from 140 to 160 kilometers per hour. In 1967 the Capitole between Paris and Toulouse reached a speed of 200, a platform for the Train à Grande Vitesse (TGV).

Automation of the railway systems progressively spread to assure coordination of signals and switches. Route settings were introduced after 1900, and centralized long-distance command systems became common as electricity gradually eliminated other means of transmission. In the 1950s Louis Armand developed a vision of railways operating on the principle of cybernetics, a program realized thanks to electronic procedures. Relay stations assured virtually instantaneous power transmission and permitted an extension of control zones. Electromechanical relays were replaced by electronic relays. In the 1980s and 1990s, computerized relay stations took charge and integrated the TGV into this research.

The first French TGV was inaugurated on the line between Paris and Lyon in 1981. It was the culmination of a program of planned research launched in 1967 within a general reorganization of the railway industry. The objective was to create a passenger service combining high speed, great frequency, a broad network with compatibility between TGV tracks and the rest of the system, and the democratization of long-distance travel. Theoretical studies conducted between 1968 and 1970 indicated that the system would be profitable because of the intense use of equipment and the possibility of climbing grades up to 3.5 percent, among other reasons.

Applied research included theoretical analyses, computer simulations, and testing. This required close cooperation with outside actors, suppliers, and study groups. Thus, the problems of stability and swaying were resolved through collaboration with various other government agencies, such as the Office National d'Études et Recherche Aérospatiales (Onera). Engineers preferred proven technologies adaptable to the new system rather than entirely new ones. The solution of a diesel turbine engine was initially adopted for the TGV in 1972. However, the Direction des Études de Traction Électrique (DETE) undertook research to adapt the TGV to electric power. The final choice of this energy source in 1974 depended on several considerations, like the nuisance of gaseous odors from turbines. Not the least of advantages was greater comfort through better coupling devices. Automation was introduced during the 1960s. After inauguration of the first TGV, a network was formed within the general French railway system, and its effects on the mobility of passengers and on the national administration were remarkable.

Interconnections: Networks of Electricity

By 1900, nearly two decades after the Paris electricity exposition of 1881, the modest helpmate of technology in the 1880s had become the main force of its evolution. The 1890s saw the creation of powerful concerns like the Allgemeine Elektrizitäts-Gesellschaft (AEG) in Germany and General Electric in the United States, both present in France.

In the beginning, electric current was essentially limited to lighting night work in large spaces like the Northern Railway Company's workshop at La Chapelle, installed in 1872. Whereas personnel increased by 30 percent, the space occupied was reduced by half. This type of large private installation had a bright future. There were 1,398 of them in France by 1907, mostly factories, railway stations, warehouses, and department stores. By contrast, household lighting with storage batteries or dynamos operated by a gas or steam engine proved to be defective, inconvenient, and unpromising.

Between 1878 and 1882 the Edison laboratory devised a lighting system with low-frequency direct current and opened a station in New York that distributed electricity within the radius of a mile. Several stations of this type were established in the United States and Europe in the 1880s. In France in 1886 Lucien Gaulard constructed a station for low-frequency alternating current, just as Elihu Thomson had done in the United States during the 1870s. France had 155 electrical stations (of which 128 used direct current) in 1888 and 400 by 1895. But the distance of distribution remained limited. At the 1881 exposition, it was acknowledged that the future of electricity depended on the ability to attain long distribution distances in order to create networks, something that was only possible by deploying high-tension lines. This touched off an intense polemic in France and the United States between partisans of alternating versus direct current. Marcel Deprez, professor at the Conservatoire National des Arts et Métiers, conducted a series of experiments on high tension in the early 1880s. The reaction in ensuing academic reports was mixed, to say the least. The engineer Ernest Boistel could speak in 1887 of "a complete ineptitude of conception" and "an insufficient study of machine construction."[5] A rival procedure using alternating current, based on a patent obtained in London by Gaulard and John Dixon Gibbs, was tested for long-distance electrical power by employing secondary generators, a process allowing a reduction of tension at each relay from 3,000 to 100 volts thanks to a transformer. Deprez's supporters stoutly contested the validity of Gaulard's method, but an experiment Gaulard conducted at Turin in September 1884 was approved by Professor Galileo Ferraris and won a prize from the Italian government. Gaulard died in 1888. His patent rights in the United States were purchased by the Westinghouse firm, which soon began installing transformers. An argument broke out between Westinghouse and Edison, a partisan of direct current, who mounted a press campaign contending that alternating current was dangerous. But a team at Westinghouse under the electrical engineer William Stanley, Jr., succeeded in improving the system of Gaulard and Gibbs. The feasibility of that system was demonstrated by two German electric companies at Frankfurt in 1891 and by Westinghouse at the Chicago exposition of 1892. Alternating high-tension current prevailed, giving rise to genuine electric networks that were much more extensive than before. The possibility of constructing high-tension lines over long distances and distributing energy infinitely ushered in the grand era of electricity, a radical break in the history of humanity. But it still remained to elaborate its uses.

At the end of the 1880s electricity was used mainly for lighting. Motors run by direct current lacked power. In 1888, however, Frank J. Sprague started the first electric urban tramway in Richmond, Virginia, using the stator electric magnet. Meanwhile, Nikola Tesla invented a motor with multiphase synchronous alternating current, and AEG perfected an asynchronous motor, opening the way for the mechanization of workshops and factories. By 1907 France counted 1,413 electrical generating plants, of which 831 were driven by water and 413 by heat, while 169 were mixed.

In 1913 7,000 out of 36,000 French communes were electrified. Distribution of current was thus "assured by a multitude of enterprises of very diverse nature and size."[6] Some of them had been founded by small local industries. But by 1894 electrical companies were starting up, and in 1906 six firms to which Paris had granted concessions since 1889 were united as the Compagnie

Parisienne de Distribution de l'Électricité (CPDE). Yet "Paris remained divided into two zones with incompatible standards," a diversity that characterized the entire French network.[7] In the provinces, big companies created ever larger central electrical plants and networks. For example, the Société Générale Force et Lumière had 1,200 kilometers of electric wires serving 107 communes as well as the Lyon tramway system.

Networks were thus created, but the diversity of their parameters made regulation of them increasingly difficult. They needed to be coordinated and safer. The answer was interconnection. In June 1920 a committee was formed to study a project to construct a national transportation network. The adopted program foresaw the creation of a "single polyvalent system for transmitting energy," organized with several lines of 100,000 volts forming five large regional networks. The work of this committee was conducted in parallel with that of a committee to study the electrification of railways. Thereby "a symbiosis of the two systems" was achieved as engineers became excited by the "mystique of interconnection."[8] It was the era of automation and telecommunication, regulation of frequencies, and telephonic dispatching. All of these advances required the utilization of powerful calculators and the success of many manufacturers.

During the 1920s eight regional groups were organized with high-tension lines that attained a maximum between 150,000 and 220,000 volts. The two most complete groups were created around railway companies: the Union des Producteurs d'Électricité des Pyrénées-Occidentales (Upepo), formed in 1922 by the Southern Railway Company under Jean-Raoul Paul; and the Paris-Orléans Company, directed by Hippolyte Parodi. Another network was provided by the interconnection between two large hydrological works in the Massif Central and heating plants in the Paris region, made possible in 1927 by a high-tension line of 90,000 volts that was increased to 220,000 volts in 1932. A dispatching center was installed in the Rue de Messine in Paris. Unlike the railway network, the electricity network, developed after 1930, was on the periphery of French territory, and the regional groups were not at first interconnected.

In June 1938 the Groupement de l'Électricité was put into place, uniting the entire electricity sector. This consolidation was completed after 1945 by Électricité de France (EDF). The total length of high-tension lines increased from 899 kilometers in 1923 to 64,000 in 1932 and 124,000 in 1945, with an operating maximum of 220,000 volts. Whereas the railway system was initially conceived as a convergence toward Paris, the electrical network was the result of moving from a local to regional situation, and from regional to national.

Mass Consumption

During the nineteenth century, the movement to create new domestic products accelerated integrating increasingly sophisticated technologies. Use of such products and services after 1830 was at first confined to the middle class. It became universal by responding to a desire for cleanliness through exacting hygiene, effective medical practice, and comfortable living quarters in which women, once freed from burdensome tasks, could rise from servitude to the status of domestic manager and gain equality with men. But this was also a response to a hedonistic wish for personal fulfillment, manifested in greater attention to personal grooming, sports, social mobility, and rather frenetic communications. To meet these needs, mass consumption and production were developed to allow the accumulation of durable goods, both individual and collective, that had to be manufactured, packaged, distributed, and destroyed or discarded after use. In addition, there appeared an abundance of various services that had to be envisioned, organized, and regulated to assure their safety for one and all. Mass marketing by large international enterprises, a major component of the so-called consumer society, was based on stirring desires, creating new needs, and standardizing modes of consumption.

After 1880 household electricity became the mainstay of mass consumption. Electrical networks brought well-being to households in the form of lighting and mechanization of family tasks. Lighting entered private space with the invention of the incandescent bulb, which triumphed over the competition from gas lamps after 1900. Electric lighting altered the rhythm of home life by abolishing the night. In 1895 one enthusiast praised "the value of fixed light, the purity of air, the suppression of extreme heat, matches, candles, oil, human well-being, art collecting, book illustrations, property, gaiety, health, and longer life."[9] At first reserved for drawing rooms, electric lighting was extended to entire apartments, thereby favoring reading. Combined with mirrors, it created a new image of the body. After 1900 and especially in the 1920s lighting was fully integrated into an architecture, whose forms, as Robert Mallet-Stevens stated in 1937, have been transformed "by bowing to the exigencies of lighting."[10]

The small electric motor rapidly conquered domestic territory. The sewing machine, one of the very first triumphs of electricity, was quickly followed by a host of varied instruments. Some, like the electric iron, the kettle, the coffeemaker, and the radiator, produced heat. Others were small machines such as electric fans, sweepers, coffee grinders, all sorts of mixers, and electric razors. These objects have been refined, but their operating principle has not changed. The electric washing machine is a good example of this mutation within continuity: it integrates a growing number of functions previously performed by hand, like washing and drying, that go beyond simply rinsing. The entire sector of household appliances has undergone a more or less complete process of mechanization and automation, first electric, then electronic.

Mass Production

Mass production was principally based on the technologies of mechanics and electromechanics, which were put into service through four technical procedures: interchangeability of parts, production on assembly lines, mechanization or automation, and new methods of industrial manufacturing.

Adopted long before by the chemical industry, assembly lines were installed at the end of the nineteenth century in light industries like food and tobacco, and then by industries working on products created by fusions or liquids, such as mirrors, glassware, and paper. In the auto industry, the full-scale construction process adopted in 1912 by the Ford factory in Detroit achieved a synthesis of three industrial practices: use of specialized machine tools, interchangeability of parts, and the assembly line. This system spread into many other sectors. In light industry, machines performing multiple functions allowed flexible production. In heavy industry, however, specialized machine tools, conveyor belts, and moveable cranes had to be coordinated. Automation was gradually introduced. The entire system required fluidity of movement and automation that was as complete as possible. These installations had a single production track whose functioning was centrally controlled. Until the 1950s these assembly chains were fragile and had frequent breakdowns, but during the 1960s microelectronics gradually replaced mechanical and electrochemical systems, imparting strength and durability to the entire process. The logic of continuous production remained dominant in industrial organization, and 570,000 French workers still worked on assembly lines in 1991.

During the 1890s electricity in workshops and factories gave a new thrust to mechanization, which until then had remained limited because of the indivisibility of energy. Electricity could be adapted just as well to small workshops as to large industry, and automation could be achieved by either electromagnetic and electrochemical processes or electronics. It was adaptable to systems combining the reception, transmission, and transduction of signals, and thus the release of a mechanical effect that replaced the action of an operator, a sequence that was actually much older than the use of electricity. Receptors could record increasingly numerous phenomena like movement, temperature, light, or a magnetic field, and transductors set off effects as diverse as the heat of a radiator, the light of a bulb, or the start of a manufacturing procedure.

Between 1949 and 1953 MIT provided the US Air Force with an automatic, digitally controlled drilling device. This was the starting point of the automatically programmed tool (APT). In 1956 the Fujitsu firm began producing a system utilizing integrated circuits, which considerably simplified operations. This procedure was rapidly applied to all sorts of machine tools: mechanical drills, boring machines, electric drills, and various others. In metallurgy the rapid cutting device developed by Frederick Winslow Taylor and the operations of casting, forging, and stamping were improved. Soldering replaced riveting, and then adhesives replaced soldering. In the automobile industry, new techniques for treating materials that emerged in the 1960s included the use of the electrophorus, infrared drying chambers for painting, and thermal applications.

The Rise of Communications

Since the end of the nineteenth century, technologies of information and communication have been adopted in increasingly rapid rhythm. In the 1960s they became ubiquitous in every realm of activity and social life as objects tuned to networks of energy, sound, or picture invaded offices, factories, and homes. Beginning in the 1970s their presence became even more absolute with the simultaneous emergence of new circuits, the eruption of the Internet, and the proliferation of cell phones. The "unconnected" henceforth constituted a new category of the excluded. To understand such a transformation it is necessary to revisit the major steps in constructing networks and the Internet, and to inquire about the consequences of these global technologies.

Texts, Images, and Sounds

The methods of reproducing texts, images, and sounds that appeared on the market in the 1880s represented the opening phase of audiovisuals in homes and in public. Especially in newspapers, a series of innovations married text and pictures. Photography thus became an indispensable presence at all public and private happenings. Instantaneous images penetrated daily life and conquered the media. Photogravure enabled the proliferation of posters, printed wrappings, catalogs, and advertisements. Even before 1914 cinema appealed to the middle class and often to lower classes. The phonograph, which Edison invented in the 1870s for dictating letters and permitting the dead to talk, became an instrument for listening to music when Emile Berliner introduced records in 1887. The phonograph industry arrived between 1895 and 1900 and quickly conquered the household. The first typewriter was sold in 1874. Its initial users were individuals like reporters, lawyers, authors, and clergymen, but its utilization in enterprises and administrations spread rapidly after 1890.

Between the world wars, and especially after World War II, these systems benefited from major improvements thanks to mechanical tools, electromechanics, and electronics. Thus, talking motion pictures came to be in the

1920s, long-playing records in 1948, and compact discs in the 1970s. The use of photocopiers, computers, and scanners improved techniques of printing. Polaroid and point-and-shoot cameras, videocassettes, and tape recorders were marketed in the 1950s, followed by the DVD and the video camera in the 1980s. Electric typewriters were commercialized in 1923, and after 1950 techniques for composing and reproducing texts developed into a huge sector of office equipment before being absorbed by computers.

Telecommunications

Between the 1870s and 1900 two identical scenarios, each in two stages, played out in the realm of telephones and radios. In the former, the first stage was the successful transfer of recent scientific discoveries in electromagnetism into the real world through the patents of Elisha Gray and Alexander Graham Bell in 1876. This enabled transmission of the voice and considerably enhanced communications compared with the telegraph. In the case of radio, the first step was the conversion of James Maxwell's theories into experiments by Heinrich Hertz in 1888, followed by the perfection of those results by Oliver Joseph Lodge in 1894 and particularly by Guglielmo Marconi in 1901. Marconi constructed a transatlantic connection and soon advanced his process by signing an exclusive contract with Lloyds of London.

The second stage in both of these developments was a major episode in the history of electronics: the invention of the diode by John Ambrose Fleming in 1905, then the triode by Lee De Forest in 1906. This miraculous device gave birth to the vacuum tube, which hugely surpassed the results of the 1900 invention of the inductance coil by the Columbia University professor Michael Pupin. The vacuum tube represented a remarkable leap forward in long-distance communications, leading to a line connecting New York and San Francisco in 1915. A merger of electronics and the wireless telegraph was meanwhile accomplished. This invention made the radio possible, and the first broadcast occurred in 1920. Vacuum tubes thereby became, until the 1950s, "the king of radio-electronics," parallel with telephonic technology.[11]

The institutional framework of telecommunications was the so-called natural monopoly. It was in the hands of either the state, as in France, or private enterprises like AT&T in the United States. Such organization went unchallenged until the 1970s. Systems of telephonic transmission by electrical current were installed in the 1920s. The first transatlantic connection by submerged cable was achieved in 1956 with the use of vacuum tubes. Their capacity increased six- to tenfold between the 1950s and the 1970s. At first telephone subscribers were connected by manual extensions called jacks. Mechanical devices were introduced in the 1920s and particularly in 1940 with the crossbar system. This did not provide a true instantaneity of verbal exchange, however, and vacuum tubes proved inadequate in this regard. Only transistors brought a satisfactory solution.

Automatic neutrodyne circuits were developed for radios in the early 1920s. The transistor, invented in 1947 by a team of physicists at Bell Laboratories during a project for improving vacuum tubes, realized its promise with the invention of interconnected transistors in 1955. This enabled conveyance of electric impulses by binary means of interruption and amplification, permitting the manufacture of semi-conductors. Thanks to the discovery of silicon in 1954, transistors gradually replaced the usual miniaturized radio tubes. In 1957 Jack St. Clair Kilby of Texas Instruments and Robert Noyce of the Fairchild enterprise invented integrated circuits that could be industrially produced by forming thousands of interconnected transistors engraved on the same silicon chip.

Progress in the technology of vacuum tubes led to the possibility of television networks. In the United States, eighty channels were on the air in May 1940. The first really operational broadcasts in France began in 1949. Color television appeared in 1964. During the 1960s wireless beams and then satellites extended broadcasting areas and internationalized them. The first telephonic and televisual connections between the United States and Europe were achieved at Pleumeur-Bodou on 11 July 1962. The first satellites could assure only part-time transmission. A valid solution was realized by putting stationary satellites into orbit, made possible by increasing the power of rockets. An international consortium, Intelsat, was created in August 1964 to regulate them. By 1971 the Intelsat network comprised 114 channels in seventy-two countries.

From Mimeographs to Computers

At the turn of the twentieth century, large public and private organizations were dealing with increasing numbers of messages. The evolution of all scientific disciplines created ever greater demand for calculation and selection of data. By the 1920s existing machines for writing, copying, and reckoning could no longer satisfy these needs. Use of calculators had expanded after 1890. In 1894 Charles Franklin Kettering invented the first electric cash register, which was rapidly and widely adopted. Mimeographing was born of the need to mechanize the American census of 1890, for which Hermann Hollerith was the only one to offer a suitable machine. Its use spread quickly in large organizations—railway companies, department stores, banks, insurance agencies, factories—thanks to the inventor's efforts to adapt it to their needs. Although very wealthy, in 1911 Hollerith sold his business, the Tabulator Machine Company, to Charles Ranlett Flint, who merged it with other office machine firms to found an enterprise that subsequently became IBM. Use of such machines took off in the 1930s and 1940s.

The technology of office equipment flourished between 1920 and 1960, leading to increasing automation and integration of functions, and thus more intense competition. Punch cards prevailed in multicopying in the 1920s, when IBM set the standard with a card of eight columns. With this new technology, multicopying became a large industry hiring a huge labor force. The first computers were manufactured in the United States between 1939 and 1945 to meet military and scientific needs. The very first one weighed five tons and utilized 175,000 electrical connections. The second, the Electronic Numerical Integrator and Computer (ENIAC), had 19,000 diodes. A decisive step came with the invention of two other computers—the Intelligent Autonomous System (IAS) and especially the Electronic Discrete Variable Automatic Computer (EDVAC)—between 1946 and 1951. The latter incorporated the ideas of John von Neumann by combining a memory and a system of programming with the calculator. Therewith the structure of computers became established for a long period. This first generation of machines using vacuum tubes developed during the 1950s made great progress.

A second generation of computers with transistors was marketed from 1959 until the 1970s. It was conceived not only to meet scientific, military, or engineering needs but also those of managers and entrepreneurs. Semi-conductors brought about a whole series of small computers adapted to the demands of large centralized enterprises. In 1964 IBM marketed a series called 360 that operated entirely within the same system. Thus, computer networks appeared as a result of multiprogramming technology. The Semi-Automatic Ground Environment (SAGE) network, designed to protect American airspace from the danger of Soviet atomic attack, featured a detection system coupled with telephonic lines. The creation in 1963 of a coding system called the American Standard Code Information Interchange (ASCII) gave a striking impetus to calculation, thereby opening the path to a general technology of numbering signals and programming languages.

The Birth of a Communications Society

In sum, the second industrial revolution, arising in the 1880s, was at first a reply to major difficulties apparent in the decade before 1870. Technological networks, especially railways, required much more effective regulation. Available energy, whether from water or steam, was neither transportable nor divisible. In workshops, loss of power was considerable and accidents frequent. The problems posed by the dispersion of workshops and by burdensome overhead expenses were becoming increasingly difficult to resolve. The absence of an efficient compact motor weakened the position of small industries against larger ones. For these reasons social stability seemed threatened.

To these technological obstacles were added expectations that only electric lighting could satisfy. The nineteenth century had lived in terror of darkness. Progress in public and private lighting from candles to gas did not bring a satisfactory solution. But electric lighting was a decisive breakthrough. The filament replaced the flame so that lighting became white and stable, imitating sunlight, which is why it triumphed despite its higher cost. It was, moreover, vigorously supported by hygienists. This applied equally well to other technologies: synthetic dyes democratized color; the synthesis of ammonia countered the threat of famine, especially in wartime Germany, due to lack of nitrate from Chile; the automobile responded to a need for mobility that neither trains nor bicycles could entirely meet.

The twentieth century was also marked by the growing control that large-scale technological networks established over society. Nineteenth-century networks like canals, railroads, the telegraph, water, and gas were joined by networks of electricity, aeronautics, telephones, and radios. The extension of these networks proceeded from technological development tied to large-scale economies and also from users' desire for increasingly distant and frequent contacts. Communication networks were at the heart of a computer technology that was the product of a fusion of copying, calculating, and telecommunication. Computers have bolstered the operation of networks and production systems. Mass society has thus gradually morphed into a communications society whose basic technologies were born of a need for regulation of these networks and systems.

Chapter 12

From Microprocessors to the Internet

Between 1970 and 1990 technological convergence produced a system of integrated networks on a worldwide scale. The proliferation of microelectronics led to the merging of communications networks, computers, global broadcasting, and audiovisual technology through constant research, adaptation to social needs, and perhaps even more importantly the power "of futures imagined by engineers."[1]

After integrated circuits were discovered and increasingly in use, the Intel engineer Ted Hoff invented the microprocessor, first marketed in 1971. It contained circuits integrated on a single silicon chip, and it performed all the basic functions of a computer. The combination of hundreds of thousands or millions of transistors allowed the execution of most of the programming functions of personal computers, electrical appliances, watches, video cameras, airplanes, locomotives, machine tools, silicon chips, and electronics. Semiconductors replaced magnetic discs for recording and storing data and made the development of microcomputers possible. Telecommunications were able to move from analog to digital transmission.

Discovery of the first laser dates from 1960, when it was but a "solution in search of a function."[2] Many possible uses were discovered and spread during the 1980s. Its light beams could carry both energy and information. Its applications were of concern for all of industry, document preservation, and telecommunications associated with fiber optics. All technologies of imaging, such as radiography and radiology, were transformed by the scanner. Sondes or antennas of all sorts were able to capture audio signals. Electronic and optical procedures came to govern the remote control of automatic systems in factories, apartment buildings, and networks. Captors were devised to simultaneously combine several types of information. They could be replaced by miniature computers capable of initiating a mechanical action immediately after analyzing transmitted data. These extreme forms of automation are used today in both highly complex engines and products for mass consumption.

Thanks to the flexibility of semi-conductors, a third generation of computers arriving at the end of the 1970s included supercomputers, microprocessors, and microcomputers. Microprocessors are hidden computers in electrical appliances, television sets, CD and DVD readers, and cameras as well as airplanes, locomotives, subway engines, etc. Interconnected, they perform increasingly complex tasks such as driverless propulsion. Born at the end of the 1960s, microcomputers were initially marketed with the Altair 8800 in 1975, but the first machine to enjoy widespread popular success was commercialized by Apple in 1977, then shortly followed by products from IBM and Compaq. By 1990 150 million personal computers had been installed worldwide, in contrast to ten million in 1980. The first real laptop, the Osborne, appeared in 1981. Compaq introduced the first compatible PC in 1985 and the first notebook in 1989. Bill Gates, who had supplied the Basic computer to Altair, founded Microsoft in 1977; it rapidly became pre-eminent in software production throughout the entire world. Considerable progress has been made in the user-friendliness of computers. The success of Microsoft depended on mastering the interface between simplicity of utilization and the unfolding of the Internet.

Flexible Production

Systems of production changed radically in the 1960s and especially in the 1970s. The model of mass production derived from the second industrial revolution had run its course. Existing technologies assured only an inadequate command of automation and continuous movement. Toward the end of the 1980s, thanks to powerful electronics, truly automated and efficiently controlled systems were achieved on the basis of accumulated experience with microprocessors, computers, and electronic optics. The development of digital control, robotics, and electronic regulation of networks allowed the programming of operations and the supervision of specific cycles without human intervention. It was possible to use external captors to alter machine programs automatically and transmit information. Microprocessors thus permitted a flexible manufacturing system (FMS).

Robots were created in the 1950s to perform tasks in hazardous or dangerous places like nuclear plants and deep water. Industrial uses followed. The earliest robots could only be programmed for a single coordinated system of production. In the 1960s Japan began using high-precision robots for making small, light objects, an invention that spread in the 1970s and 1980s. To assure a sequence of operations, wired circuits with diodes or transistors were used to process information. Breakdowns were frequent until relay modules were replaced by electronic calculators that were programmable automatons. This was the first step in the invention of computer-controlled robots.

This technology was enhanced by the appearance of microprocessors, plus RAM and ROM computer memories, also developed in Japan in the 1990s and subsequently spread throughout industry. External captors, capable of receiving signals imperceptible to humans and interpreting the information received, enabled these automatic devices to adapt to their environment. The computer captured, stored, and processed signals, then communicated this information in a form appropriate for machines.

To offset a lack of coordination among various tools, complete digital systems of command and control were invented, including digital regulators that could manage several variables: communications between humans and machines via screens or keyboards, digital networks of interactions with robots, and outside sources of information. Such a system could be regularly adapted to new procedures and to diverse manufacturing needs. A first attempt was made in 1967 in the United States, but these installations were not actually developed until the end of the 1970s, when a common language between different sectors was elaborated and regularized. In the end the robotics industry succeeded in managing productive systems that were both complex and dangerous, such as aluminum factories and nuclear plants, thereby creating flexible mass production by furnishing varied and virtually flawless objects. Such performance allowed a specialization of markets without sacrificing massive automated systems.

Telecommunications

The capacity of telephonic lines continued to grow after the 1960s thanks to the use of transistors in coaxial cables and fiber optics in the 1970s. Systems called Modulation par Impulsion et Codage (MIC) became operational in the 1960s with the spread of semi-conductors. In France, CNET successfully experimented in 1970 with electronic components for communication. Telephone centers equipped with computers and integrated circuits soon became completely digital, and a general European standard of thirty-two channels was adopted in 1969 with relays by networks of satellites.

Public and private monopolies disappeared more or less rapidly after 1970, and infrastructures were managed by enterprises offering their services to networks. But this connection between the management of infrastructures and the exploitation of networks was gradually severed. Instead, information highways were programmed to transmit all forms of knowledge and exchange on the same network: data banks, television programs, telephone messages. The Apollo project launched at the end of the 1990s utilized a cable across the Atlantic with four pairs of fiber optics the thickness of a hair. Satellite networks, either governmental or private, were put into service. The European Space Agency was founded in 1972, and the Ariane rocket program began in 1980. Hertzian networks competed in efficiency with cables, as the Asymmetric Digital Subscriber Line (ADSL) demonstrated in 1997.

Procedures in the transmission and communication by packages of digital data became commonplace, and telephone networks, thanks to modems invented in the 1970s, assured transmission of data between microcomputers. The Transpac system, introduced in France in 1975, used a telecommunications protocol called X.25, which was standardized in 1976. In this process each package was governed by a computer. In a competing system that subsequently appeared in the United States, a network of computer relays determined the route of each package. The Réseau Numérique à Intégration de Services (RNIS), transmitting all kinds of data, was created in France during the 1980s, with competition between public and private networks. Cooperation was finally achieved between telecommunication companies and computer networks.

Early use was made of walkie-talkies on sites such as public works projects. The first prototype of a cellular phone dates from 1973. But analog cell phones were cumbersome and unreliable. The change to digital apparatuses offered more potential. The cell phone did not really develop until 1984. In 1987 the Conférence Européenne des Postes et Télécommunications (CEPT) defined a common standard called the Global System for Mobile Communications (GSM) that was first implemented in 1992. For the first time, a single network of European telecommunications was programmed. Television transmissions soon followed.

Information and Audiovisual Technologies

Between the 1960s and the twenty-first century, information and audiovisual products available to enterprises and households improved considerably, and the Internet "unified all previous media that it received and diffused."[3] Communication networks created around the family a coherent technological environment composed of four types of apparatus: those that functioned alone, like the electronic camera or the calculator; those that depended on outside software, such as personal computers, magnetic phones, video cameras, and compact or video discs; those that depended on centralized networks, like television, radio, and the telephone; and those dependent on a web of telecommunications like the Internet, which increasingly prevailed as the universal instrument of communication and information.

The invention of these products of mass consumption required semi-conductors and very precise techniques of digitalization, optics, and mechanics. Manufacturers had to define norms and standards permitting the interchangeability of products and software. In the case of the audio-digital compact disc perfected in 1969, for example, standardization was the result of an accord reached at the outset of the 1980s among several firms, including Philips and Sony. The emergence of DVD technology in the early 1990s caused a replacement of the videocassette, a process now completed. Other devices were dependent on telephone, radio, or television networks.

The growth of telecommunications radically altered the status of radio. Beginning in the 1980s the use of FM spread to increasingly numerous and specialized stations. This trend reached television with the advent of cable networks, satellites, and the total or partial deregulation of audiovisual. By the end of the 1980s a great majority of households had color television. Flat screens and high-definition TV considerably improved picture quality. The rise of digital television was less rapid than expected, but once in place, the television set became technologically comparable to the computer by allowing the reception and diffusion of myriad data and images, cable or not.

Computer Networks and the Birth of the Internet

Computer networks developed willy-nilly in the 1970s. Often informal, they were intended to be administered cooperatively with a minimum of central coordination. They frequently used software that manufacturers had developed for other purposes. In the United States, Bell Laboratories perfected the Unix system, which allowed construction of compatible information tools sharing a central computer equipped with a modem. The 1978 version of the Unix thus allowed data to be exchanged via telephone lines. Having a computer only made sense if it was connected to other computers. Several university networks chose Unix and created an informal electronic message service called Unix-to-Unix Copy Protocol (UUCP). Other networks were created by firms like IBM and its clients or by specialized companies such as Compuserve and Prodigy. Others, for example AT&T and specialized companies like Novell and Cisco, sold ready-made networks to administrations or enterprises. But users of often-incompatible networks wanted to break out. Only the largest possible connection would thus fully reward computer investments.

At the origin of the Internet in the United States was the creation in 1958 of the Advanced Research Project Agency (ARPA), designated to solicit and incorporate university research useful for national defense. In 1971 this agency installed a computer network uniting all the research centers working for it. It selected a means of communication by packages having their own protocol. The use of e-mails exploded. To resolve the problem of connecting one network with others, the Network Working Group chose to organize an interface among heterogeneous networks rather than following the contrary principle of unified connections adopted by the French university network Cyclade, designed by Louis Pouzin. The Transfer Control Protocol/Internet Protocol (TCP/IP) was ready in 1978. This common language allowed all networks to communicate and furnished them with a uniform method of address, thereby eventually building a worldwide system.

The proposal called X.25, intended to furnish the basis for gradually constructing a universal network, was adopted by numerous governments and several private networks like IBM and Digital Equipment. But this effort was weakened by a decision of the Federal Communications Commission (FCC) in the United States that obliged AT&T to abandon the computer market. Moreover, the adaptation of X.25 proved so difficult that the International Standardization Organization (ISO) left open the possibility of accepting other proposals.

In the 1980s the internal dynamic of the TCP/IP system demonstrated its viability. The Internet combined existing networks into a complex based on the compatibility of different technologies. By the end of the 1980s several million computers could exchange e-mails, and the different networks began to establish internal connections. A group of Internet users created in Europe in 1989 comprised four hundred organizations by 1996. The Internet Engineering Task Force (IETF) was created to deal with operational difficulties. Until then, use of the Internet had been limited to e-mails and file transfers. But in 1990 a team at CERN conceived the Web, defined as a human pool of knowledge that facilitated sharing of multimedia data. This project transformed the Internet from a research tool into a popular medium capable of attractive and profitable functions. This network of networks thus became the instrument of a dual revolution: a new economy based on an intensifying exchange of information, and the outburst of a multimedia civilization. During the 1980s search engines such as Archie and Gopher appeared, supplying series of keywords to enable exploration of the living encyclopedia that the Internet became.

Communications of a Large Network: The SNCF

The role telecommunications played from the 1960s to the 1980s in the operations of the French national railway system, the Société Nationale des Chemins de Fer (SNCF), serves to illustrate the integration of large technological networks and communication networks into a unified system. In 1968 the SNCF inaugurated one of the first heterogeneous telecommunication systems in France. The increasing number of its uses during the 1970s necessitated a more powerful network to transmit data. Thus was born the Réseau de Téléinformatique Ferroviaire (RETIF), which in 1984 became the Réseau de Téléinformatique à Commutation de Paquets (Retipac), to which the various computers of the SNCF were connected. By 1990, this network possessed 6,200 terminals. The development of these networks increased the use of microcomputers. The SNCF purchased 1,500 of them from Logafax in 1978, and in 1984 it adopted the IBM/PC. Its computers numbered over 10,000 by 1990.

The first operation justifying the study of a telecommunication network, in the 1960s, was the Gestion Centralisée du Trafic Marchandises (GCTM). This system, completed in 1979, was developed to replace several superannuated programs such as the Nouvel Acheminement des Wagons (NAW), customer service (SESAME), dispatching (MARS), and maintenance (EDI Fret). A system for train reservations (RESA) began in 1973 and was expanded throughout the rail system in 1974. A project for travel offices started in 1972 grew into a vast system connected with Air France, which by 1983 included nine hundred agencies. The automation of billing, begun in 1968, resulted in a payment network attained through an accord with large banks. At the beginning of the 1990s, foreseeing a boom in reservations and transactions brought on by the TGV, the SNCF acquired a license from the computer firm that American airline companies used for reservations, Sabre, whose development took ten years in the 1950s and 1960s. Its launch was difficult, since it required a pricing reform. Other computer uses in the SNCF, as in all industrial enterprises, concerned administration, production, and research.

The Second Life of Networks, 1995–2008

In the 1990s the Internet became "the great worldwide and multilingual reservoir of services and information."[4] Its accelerated performance after mid-decade was brought about by implementing research and navigation tools, using high-intensity lines, and lowering service costs. The real starting point in the use of global positioning was Netscape Navigator in 1995. By the following year it had 84 percent of the market for Internet searches. Meanwhile, Microsoft offered Internet Explorer, marketed as a complement to Windows. A struggle ensued, marked by a lawsuit against Microsoft that finally resulted in a compromise. Yet Microsoft controlled 81 percent of the market by 2007. Access to the Internet was subsequently extended to new terminals, notably cellular phones, thanks to mobile high-intensity lines.

The connection between computers and telephones was achieved by using modems. In 1995 the intensity of these lines did not exceed 56 Kbit/s. By 2005 this had been multiplied by a factor of 150 due to the use of ADSL technology in these networks and a corresponding improvement of cable networks. The high intensity associated with cell phones had the consequence of an unprecedented surge of the Internet. The number of sites grew from 20,000 in 1995 to 160 million by 2008, "a glide from occasional Internet to an Internet perpetually in use."[5] Uses such as online banking and social networking developed rapidly, and old practices were swept away. The decline of lending fees was an outcome of exponential growth in the capacity of processors and storage of information bundles. The density of data on a disc went from 200 bits in 1955 to 100 Gbit—that is, one hundred billion bits—in 2008.

Chapter 13

Information Technologies and Society

In view of the radical and universal nature of the Internet, Manuel Castells has remarked that its "technological transformation" was comparable to the invention of the alphabet in Greece about 700 B.C. The interaction between the real world and the virtual world of the Internet could only extend the previous influence of new information technologies. In analyzing various aspects of the Internet's influence and impact on society, three factors are paramount: the evolution of the structure and strategies of enterprises, the development of cultural practices associated with the intrusion of telecommunications into households, and the changes occurring in the realm of knowledge and information.[1]

Enterprises

A strong complementarity has been established between both communication and information technologies and the organization of enterprises, either because procedures had to be adapted to the structure of enterprise or because that structure changed with the evolution of technology. Thus, three models of enterprise appeared in succession: a vertical centralized system (1880–1960), the emergence of networks (1970–1990), and the networking enterprise (1990s).

1880–1960: The Vertical Centralized System

Between 1880 and 1960 a model of vertical integration predominated, reaching its apex in the 1950s and 1960s. It was within this framework that business firms adopted the bureaucratic tools apt to assure functions of operation and management, including typewriters, calculators, copy machines, and early on in the United States, multicopying, which also conquered Europe in the 1920s and peaked in the 1950s. These inventions served to do invoicing, follow the stock market, keep the books, assure the payroll, organize the files, and so forth. But they were increasingly ill suited to the evolution of business affairs. Beginning in the 1950s computers became a mass market. The first machines used punch cards and slowly adapted to multicopying. Computers and the first telecommunications networks buttressed the vertical structure of large centralized enterprises. The very conception of computers based on a hierarchy of data banks reflected this type of organization. This was a response to the imperfections in the circulation of knowledge, precisely at a moment when company managers wanted to reduce the uncertainties of research and to better control its results. As a consequence, computers very soon became the main components of automated productive systems.

1970–1980: The Emergence of Networks

The model of vertical centralization entered a crisis in the 1970s and 1980s. Its effectiveness was cast into doubt, in regard to both the capacity of productive systems to adapt to increasingly diverse fluctuating demands, and the validity of research systems organized in large laboratories. The discovery of new products comparable to nylon seemed increasingly rare within these large organizations, whereas such products were being developed in smaller enterprises and by Japanese firms where research was not centralized but carried out in situ.

In the 1990s the utilization of microcomputers spread to offices and factories. Seven million of these devices were installed in American enterprises between 1981 and 1985. They became the preferred terminals for complete networks and instruments of their transformation into multimedia. From the 1960s to the 1980s builders developed a series of specific applications. Printing of texts spawned computer-assisted preparation of images and of publications. Computers revolutionized retail commerce. Point-of-sale techniques allowed assessment of sales and stocks in real time and thus enabled last-minute purchases of products and permitted better management of merchandise. Automatic checkout of products appeared in supermarkets toward the end of the 1970s, and by the beginning of the 1980s all products were coded. In industry, electronic tools and computers permitted a response to the evolution of more varied markets. The demand for variety required a complete revision of Taylorism and Fordism as models. Projects became collective efforts in which many actors participated from conception to production. Such programs depended on the use of computers in industries from clothing to automobiles. Hence the advent of microcomputers rapidly brought reorganization to the networks, bringing down barriers separating upper from middle management and favoring the adoption of decentralized operations adapted to markets' tendency toward greater diversity and flexibility. This created numerous sectors of a new type of enterprise founded on cooperative networks of commercial activity in production and research, both within enterprises and in their relationship with the outside world.

The 1990s: The Networking Enterprise

An enterprise adopts a networks-centered strategy in two steps. First, e-commerce requires a new organization of commercial relationships founded on yield management in both the definition of products and in pricing and marketing policy. Second, e-management requires a new organization of the enterprise, its relations with outside partners, and its reckoning of real costs.[2] The first step is supposed to exercise a beneficial influence on sales. The designer of a business site should "put himself into the head and skin of a customer" by establishing a dialogue.[3] Yield management varies the offer of a product at different prices according to the nature of its clientele and the competition of marketing strategies. E-business makes a genuine negotiation possible between enterprises and clients, who are then able to find the product best suited to their needs.

Both inside and outside of enterprises, the Internet radically alters the nature of relationships between different functions and various actors. These relationships form, beyond any hierarchical logic, around connections assuring a collaboration among the decentralized components of the system. This arrives at a definition of objectives and programs that are shared by everyone but can be constantly reshaped. The performance of a network also depends on its coherence. This model of organization has the advantage of orienting the activities of an enterprise toward innovation, becoming "a key weapon of competition" insofar as it has the capacity to treat information in real time and adapt itself to "the variable geometry of the global economy."[4]

The technology of networks exercises direct influence on industrial and financial strategies and, more generally, on decisions of operators in capital markets. The telephone was a powerful instrument in the globalization of financial markets and products in the nineteenth century. Various financial markets follow their own specific rules, but connections are established among them thanks to the invention of products or the adaptation to regulations. A global market is thereby created in which financial networks function in close relationship with increasingly numerous information and communication networks that permanently and instantaneously broadcast news. These possibilities make them virtually uncontrollable and encourage a vast bull market.

Objects of Daily Life

The most general model for the diffusion of household items was initially a slow progression among the well-to-do, followed by an acceleration through imitations within the middle class, and finally a spread to the lower classes resulting from lower prices. This scheme was applicable, for instance, to the refrigerator. In 1965 95 percent of upper-class families in France were equipped with refrigerators, compared to only 50 percent of the general population. By 1984 90 percent of all households had a refrigerator. That pattern was somewhat different for television sets, which underwent rapid development with little difference among social categories. By 1981 the spread of TV was complete. The same pattern held for color television. The model of washing machines approximated that of refrigerators and televisions with a total equipment of families with children by 1982, whereas dishwashers—at 8 percent in 1974 and 18 percent by 1982—remained a luxury. Hence the three products usually owned by all classes were the refrigerator, the television set, and the washing machine.

During the 1970s France still lagged in household amenities in some respects. In 1968 half of dwellings had no hot water, shower, or washing machine, and only half owned an automobile. This delay was made up in the 1980s: by 1988 95 percent of homes owned a television set, 96 percent a refrigerator, 84 percent a washing machine, and 75 percent an auto. From 1960 to 2005 the lower classes' share in consumption rose from 30 to 50 percent. The technology of everyday life spread into new sectors such as food. Conservation of foodstuffs and preparation of meals were entirely changed by deep-freezing, soon in competition with new methods of wrapping. The use of microwave ovens was more frequent after 1986, leaving the impression that it was above all "a machine to gain time."[5]

A critical reaction against mass consumption gathered steam in the mid-1960s. Attempts were made to distinguish markets according to region, sex, class, age, and other criteria. This challenge could not fail to effect a progressive change in the strategy of enterprises, which henceforth devoted research to worldwide variety. Emphasis was therefore on the masculinity and femininity of products, and publicity campaigns were accordingly targeted specifically at, say, men aged 15 to 25 or women aged 40 to 60. Within this framework there developed a hedonism that was translated by increasingly diversified and often ephemeral patterns of consumption.

The Automobile

Autos were the cornerstone of a new civilization of rapid and long-distance individual transportation. The aristocratic and bourgeois products that lasted until the 1950s gradually came to include the people's car, which had emerged before World War I with the Ford Model T. But it evolved. The president of General Motors in the 1920s, Alfred P. Sloan, replaced the uniformity of Ford Motors with a strategy of variety in models and prices. Europeans adopted this model after World War II. Thus, a new era of diversity and innovation opened in Europe, founded on the integration into autos of all the possibilities that technological change offered in the way of materials and electronics.

At their outset, automobiles satisfied an aspiration for individual and family mobility that neither trains nor bicycles could meet. Between the two world wars autos spread more rapidly in the United States than in Europe or Japan due to a notable difference in the standard of living. But that delay was overcome after 1945: the number of cars in Europe increased tenfold between 1950 and 1974. In 1970 the United States counted 430 autos per thousand inhabitants, and France 240. By 2005 France had 480 per thousand, about the European median, although below Germany and Britain. The French household rate of auto possession rose from 30 percent in 1960 to 60 percent in 1973, and to 80 percent in 2006. The number of two-car families was 21 percent in 1985 and 33 percent by 2006.

The automobile substantially altered the conditions of family life by facilitating travel from home to work and also short-range and pleasure trips. Moreover, it favored tourism free of the constraints of railway travel. Each country created its own particular automobile culture. Europeans remained attached to a sporty style utilizing the driver's know-how, whereas Americans especially sought comfort, safety, and ease of driving and therefore more easily accepted automation. The oil crisis of the 1960s strengthened a current of anti-auto hostility, but in fact it did not at all alter the intimate relationship between humans and cars, even less so when manufacturers in the 1980s and 1990s improved their products' technical performance, safety, and design.[6]

Communication Devices

At the time of its appearance, the telephone was considered a strictly utilitarian instrument of communication, but it soon became a way to reduce loneliness and anxiety through conversation. Especially for women, telephones created a new form of sociability. Development was more rapid in the United States than in Europe. In 1913 Americans exchanged 14.5 billion telephone calls versus 5.8 billion by the more populous Europeans. Backward in this regard, France did not make up ground between the wars, although the number of telephones doubled to a million between 1924 and 1928. This lag was still greater after 1956, despite this figure doubling again between 1933 and 1955, and once more between 1955 and 1968. The wait for installation of a telephone was up to three years. Telephone use took off in the 1970s.

Radios at first concerned only "some passionate amateurs,"[7] but by 1930 they had gained a place in homes. There were a million radios in France in 1932 and five million by 1939, more than the number of telephones. More than half of programs were devoted to music. The radio was also an instrument for promoting sports, and after 1932 it became a major factor in political debates. The introduction of FM considerably improved the quality of programs. Automobile radios became common, and soon there was more than one radio per citizen.

Television burgeoned in the United States after 1945. By 1955 85 percent of families had a TV set, and users averaged five hours of programs each day. France made up its initial deficit after 1955, counting 1.9 million black-and-white sets in 1960. By 1975 two-thirds of the French watched television every day—including at mealtimes, for half of them—for an average of fifteen hours per week. A surge occurred in the 1960s. Between 1960 and 2008 the purchase of communication devices grew more rapidly than total consumption. The annual share of public consumption of electronic equipment expanded from 1.1 percent to 2.1 percent. While consumption of communication services quintupled, growing by 8.1 percent annually, the proportion of television sets declined from 88 percent to 40 percent, whereas computers rose from 0.9 percent to 39.3 percent. As a share of total communication expenditures, the part of telecommunications connected with the Internet grew from 45.5 to 75.8 percent, while that of the postal service shrank from 25 to 6 percent, and radio and television from 29.5 to 18.3 percent.

After 1960 household equipment with communication devices unfolded in three phases: during the 1960s the main thrust of communication expenditures was the development of telephone networks in a French nation that clearly lagged behind others; the 1960s and 1970s were the era of black-and-white television, stereo systems, and video cameras; and in the 1980s came computers, cellular phones, and then flat-screen television sets in the 1990s. Since the mid-1990s the major engines of growth have been the cell phone and the Internet, products that have become ever more indispensable. The merger of these two communication networks into one has caused all functions of leisure and communications to be reshaped into the same object. An old dream of engineers has thus been realized: every home and office is given the possibility of "gaining access with a single magic box to a potentially unlimited number of sources, words, texts, images, data, and leisure activities."

Actually, however, the premier medium remains television. Watching television "is the activity most shared by the population, the most frequent, and the most consistent: 39 percent of leisure time of the French in 2006, versus 24 percent for hobbies, 4 percent for gardening, and 3 percent for reading."[8] Even more than radio, it is a mass medium. "It is the only medium that thrills an entire country, disturbs it, instructs it, makes it laugh and cry."[9] Thus, today we are still "the children of television."[10]

The Era of the Internet and Cellular Phones

The integration of different modes of communication into an interactive global network is the result of the exponential multiplication of communication devices. Today, the world's users of the Internet number more than a billion and a half, having quadrupled since the year 2000. A billion personal computers were connected to networks in 2008. An estimated 3.3 billion mobile phones are in use worldwide, versus 1.3 billion landlines. Touch screens like that of the iPhone have gradually replaced little keyboards. There were 162 million GPS apparatuses in 2007. Video and computer games were increasingly connected to networks, and an estimated 277 million players were online, their number is rising at a rate of 20 percent annually. At this rate all the devices in a home will soon be interconnected and commanded from a distance.

Western people now spend more than a third of their time in front of a television or computer screen. Yet the Internet retains the advantage of "bringing together all previous media that it receives and transmits."[11] As this conquest began and spread in the 1990s, its promoters delivered a messianic and utopian message, rather like that of railway partisans in the 1840s. Its principal virtue, they said, was interaction: the Internet should liberate the individual from the oppression of media diffused by centralized networks, as an instrument of dialogue it will be a powerful factor of exchange and of peace, and it will allow a multiplication of diasporas and the formation of real communities that are sometimes ephemeral but may also create permanent social bonds. Meanwhile, the Internet has also been described as an evil force, destructive of human identity and all social values, imprisoning individuals within a network.

However, the Internet has only extended previous practices. In this perspective it is a potent instrument to facilitate the daily life of families and to encompass all aspects of culture and learning. It is also a creator of social ties throughout the entire world. E-commerce is slowly gaining a place in every sector of activity, either because existing enterprises open an outlet or because they exclusively practice online business. During the 1980s and 1990s the Internet thereby became a vast supermarket of all sorts of products and services meeting every imaginable need: shopping, managing bank accounts, finding an apartment, taking care of health, organizing trips, visiting museums, reserving theater tickets, gathering information, communicating abroad. In this sense it may be understood as an extraordinary means of freeing time for cultural and recreational activities or for social relationships. It becomes the opposite of imprisonment and a means of reaching out to other cultures. The utopias of its founders thus seem to be realized.

The Nature of Messages

The Internet is not just a huge assembly of data. As a place for gathering written, audio, and visual symbols, it is incomparable to any previous experience. Today there are more than 180 million sites, that is, 2.5 times more than in 2005. Google's index has 100 billion pages. Together YouTube and Dailymotion have 130 million videos, "the equivalent of five hundred years of non-stop viewing."[12] And the size of the Web reserved for institutions is five hundred times greater than that of the public Web. The immensity of accumulated data is but one of the characteristics of this network of networks. Thanks to the digitalization of its information, the Internet has become an immense library and a gigantic global archive available to everyone. The mass of texts, sounds, and images transmitted and duplicated for eternity continues to increase exponentially. Collective memory thereby becomes uncontrollable and is constructed "without limits."[13] It is not formed of separate elements but of a constant flow, and it is no longer naturally hierarchical. Hence the private life of individuals is unprotected, and still less so if they leave traces on the Internet.

Contrary to the liberating utopia imagined as the outcome of a unified system of communications stand the all too obvious distortions of its use. Besides enabling the expansion of culture and scientific knowledge, it permits the proliferation of messages of violence, pornography, and vulgarity. According to Emmanuel Hoog, the Internet "is a new tribune for all sorts of scoundrels, allowing everybody to spread his own truth as he sees fit. A neutral space, it allows the true and the false, the real and the imaginary, to coexist." He adds: "In the name of individual freedom [and] the right to information … the diffusion of the most depraved contents, the least authentic, the least analyzed, is encouraged."[14] The Internet has thus been diverted from the civilizing mission foreseen by its promoters. The urge for its actors to highlight or fabricate sensational, shocking events falsifies the very notion of information. The Western world is subjected to a continuous flow of images broadcast in real time and to information that encourages sensationalism and necessarily simplifies the contents of messages. In order to survive, the newspaper press has had to adopt a concept of information more preoccupied with sensation than with thorough analysis. In the final analysis, the Internet becomes a major instrument in the destruction of local identities and cultures, just as railways were accused of doing in the 1850s.

Social Connections

From its beginning, the telephone created informal social ties that could become enduring. It was and remains an effective instrument against isolation and anxiety, and also against cruelty, as in the admirable *Voix humaine* of Jean Cocteau. Early attempts were made to formalize these relationships via, for example, the citizens band that operated in the United States and in France, permitting discussions and contacts by persons on the same radio frequency. Likewise, in France in the 1980s, discussion groups were organized on Minitel.

In the opening phase of the Internet, like-minded groups formed using electronic messaging. If the number of participants became excessive, the system broadcast news classified by themes. Some of these newsgroups became "a journal having correspondents around the world, a journal in which you could insert your own commentaries."[15] More convivial methods of dialogue were perfected in the early 1990s. Sounds and images could be exchanged, and procedures for swapping information data were simplified by the use of search engines. Pioneers of these communities such as the Wall, Tripod, and Yahoo! GeoCities appeared between 1992 and 1995. In 2002 and 2003, a new generation of social networks surged onto the market, including Skyrock, Windows Line, and Copains. MySpace and Facebook debuted in 2004 and enjoyed great success. The ease of using programs thus became such that they "mutated into an omnipresence and ubiquity of social networks."[16] In 2009 Facebook counted 350 million subscribers in the world and 15 million in France. Half of them—not only young persons but also adults—were connected at least once a day. Blogs and social network sites have invaded every area of activity, including politics and private life, and they are increasingly affected by publicity. We have moved "from a situation where attention was plentiful and contents rare to a world where attention is rare and contents plentiful."[17] Among American adolescents in 2006, 35 percent of communications were in person and 47 percent via online social networks. The depersonalization of contacts has thus become a reality.

The use of these networks has only reinforced the trend toward various types of groups that might serve fanaticism, hatred, and violence. Networks thereby favor the emergence of a certain "social disjunction," to use Hoog's phrase. The individual's identification with a collectivity is undone to the benefit of communities gathered by sites where every possible fanaticism is expressed, which also has the inconvenience of depriving persons of ownership of their own message. Exposure on the net of private lives makes their protection impossible. The development of video networks like YouTube accentuates these perverse effects still more with their worldwide success. Thus, 41 percent of Internet users in the United States in 2007 had visited a site of this type, where very often "sordid and obscene images" were exchanged.[18]

In view of the serious corruption wrought by perverse users of information technology and the Internet, it is urgent to find a response capable of mastering it. The liberating effects of these technologies must be set against individuals' alienating dependence on them. Such dependence on large technological systems is not irrevocable.

The Unforeseen Outcome

The dynamics of technological change examined here, being the result of a dialogue between consumer desires and technological opportunities, appear rather chaotic. Along the way there are false starts, costly deadlocks, or else miracles. Society's needs are not always clear, and the responses to them are uncertain and often inappropriate. The capacity of the technological system is often overestimated and its reply unsuitable. Achieving coherence between consumer demands and technological possibility requires time, and a long period of maturation is sometimes necessary. Electricity networks and railway signals, for instance, needed much time to find proper responses to systemic imperfections. Lasers too were a discovery whose effective utilization required many years to become operational.

Technological change is unforeseen. The response of households to new services or products is usually unpredictable, despite efforts of marketing. The use of a new product might take a course that its inventor or promoter had not anticipated, as shown by the history of the bicycle, the telephone, or the Internet. In reality, the means of appropriating technology are several, as numerous illustrations have indicated. Besides, long-term choices depend on how the technological and scientific communities with the power of decision orient their research to a vision of the future, on their perception of the priorities or opportunities offered by the state of knowledge, and also on the pressures exerted by the state and by public opinion.

The chaotic sequence of technological change has not impeded the birth of a planetary system of interdependent networks around which all human activities are organized, a situation that is the culmination of a process of universal interconnection. The dependence of transportation and communication networks on electricity and radio-electric networks is total. Markets for primary materials and finance are all connected by the Internet. The daily operations of large administrative systems are subject to the procedures of the network of networks. For decades, and long before the Internet, scientific research has been organized around communication networks. Railway and aeronautic safety, like that in large industrial systems, is assured by specific networks. In many countries, households are largely sites of convergence of a multitude of integrated networks, if not the Internet. Hence most of humanity has eyes fixed on a screen during much of the day. The cell phone has given new impetus to this collective fascination, a radical departure that was not foreseeable.

Conclusion

Since the twelfth century, four types of actors belonging to the sphere of technical knowledge have maintained constantly evolving relationships: artisans and workers, experts and engineers, entrepreneurs, and intellectuals and academics. These relationships were organized according to the modalities peculiar to each of the periods identified, and they culminated with the formation of knowledge that was both remarkable in its content and unified in its methods.

Five processes have combined in the course of this mutation, whose rhythm of development accelerated after the first industrial revolution of 1760 to 1830:

1. To the extent that enterprises developed their activities, a model of centralized organization of research was gradually installed and remained dominant until the 1960s.
2. The field of scientific learning was broadened, and the constant filtering of knowledge did not hinder growth in the number of technical disciplines and currents. The resulting research simultaneously assumed an increasingly interdisciplinary character, the number of research centers rose considerably after 1850, and various forms of collective or cooperative research developed.
3. The circulation of information and of ideas became more rapid and massive.
4. The dialogue between technological and scientific research intensified.
5. Scientific activity has been increasingly integrated into society under diverse political and social circumstances.

The integration of research into enterprises was gradually achieved during the nineteenth century and experienced a decisive takeoff in the 1880s. Large research laboratories were created in the German and American chemical and electrical industries, which often exercised a monopoly. They began rather slowly but became huge research centers, their large size serving as an impetus for burgeoning economies and giant enterprises that seemed the proper response for global research in a competitive world. Through diversification, this model spread into different sectors and countries over the twentieth century. The laboratories of I.G. Farben in interwar Germany and the postwar Bell Laboratories in the United States are the two most notable examples of this research strategy, which reached its peak in the 1950s and 1960s. Beginning in the 1970s decentralized models of organization gradually developed, which gave enterprises the configuration of a network comparable to long-standing relations with outside partners. This evolution was compatible with the orientation of new information technologies. Nonetheless, the centralized model did not entirely disappear and became adapted to certain sectors.

The differentiation of disciplines and the resulting necessary interdisciplinary approach of research were not new, as the case of nineteenth-century organic chemistry well illustrates. Its integration with other chemical disciplines such as biochemistry, physical chemistry, and colloidal science was basic to its development in the nineteenth century. The great research projects of the twentieth century, like the synthesis of ammonia, could not have been achieved without parsing the learning from several theoretical scientific disciplines with knowledge borrowed from the engineering sciences. The post–World War II treatment of the massive quantitative data and complex mathematical models included in pure science could not have been accomplished without the first computers. These large programs required the mobilization of considerable financial resources that could only be amassed within the framework of collective research and public funding.

The appearance of various forms of collaborative research marked the culmination of the evolution unleashed in the mid-nineteenth century. They initially concerned the establishment of connections between enterprises and laboratories in order to promote localized research programs, either to coordinate complementary skills or to realize interdisciplinary efforts. In the twentieth century, these programs reached gigantic proportions by assembling much greater numbers of enterprises. That was the case for a consortium of petroleum companies that clustered around Standard Oil in the United States in the 1930s to invent a procedure for catalytic cracking, mobilizing a thousand persons to compete with the procedure developed by Eugène Houdry. Such cooperation could also take the form of cooperative research within the framework of a profession. Around that same time in Great Britain, the laboratories of the Calico Printers Federation thus perfected several synthetic cloths, like terylene. A further step was taken in the 1950s, especially in Germany, with the creation of state-supported research networks. In the United States this work was developed by universities like Stanford and MIT, and in Europe through research programs organized by regions, states, or the European Union. Moreover, large enterprises often sought accords or consortiums to promote programs of research and manufacturing precisely designed so as to reduce costs, as often occurred particularly in the electricity and automobile industries through mergers.

The circulation of knowledge accelerated prodigiously in the nineteenth century with the rise of transportation and communication networks that encouraged exchanges and confrontations among intellectuals. A good example of this process was the Karlsruhe Congress of 1860, which marked a major turning point in the history of chemistry from both a social standpoint, with the creation of an international chemical community, and in scientific perspective, via consensus on the definition of basic concepts. This congress was followed by the first electrical exposition of Paris in 1881, which gave rise to a definition of standards. Meetings of this fundamental type occurred in all disciplines in addition to regular congresses. Communication networks transmitting data confirmed the positive effect of these exchanges. The beginnings of the Internet, to repeat, took place within an American inter-university network similar to those that had been founded in Europe.

Hence the centralized administration of research was not, at first, hindered. To the contrary, various forms of information and idea exchange among intellectual and technicians were developed. These informal networks were based on very old practices that had been the norm in the seventeenth and eighteenth centuries. The relationships between colorists and chemists in the dye industry and between manufacturers of steam engines and intellectuals in thermodynamics are good examples. These informal relations prefigured the formation and development of inter-enterprise and inter-laboratory research networks, institutionalized or not. Such networks were also able to integrate research centers beyond universities and enterprises. They became numerous in France and Great Britain in the mid-nineteenth century, thereby extending the trend of times past.

Scientific scholars maintained a permanent dialogue with possessors of technological know-how. The proud boast of some nineteenth-century intellectuals that they were only humble servants of an emerging pure science was as dubious as the idea of a total dependence of technological learning on science, reducing it to an "applied science." Technological knowledge was constructed on the basis of a permanent contact between quotidian experience acquired in the manufacture and use of technical objects, and available scientific knowledge. It was thus that mechanics, from the nineteenth century until its extension into mass production, very early integrated scientific learning, including theoretical science. Meanwhile, the history of water power pitted a pragmatic American approach against a French procedure founded on scientific theory. The two eventually merged. It was the same for the durability of materials. Yet scientific knowledge could lead to the birth of new practices and technological pathways, as was the case in the physics of high temperature, electronics, and lasers. Utility was always an integral part of scientific projects, just as technology was forever spinning off polemics and knowledge that were part of the evolution of science.

Also dubious is the preemptory assertion by many authors that the activity of engineers and intellectuals is always subordinate to economic and political interests. The reality is simpler: scientists' relationship to active social forces is natural, the gravity of social and political factors notwithstanding. The intellectual world was submerged into society very early on. Engineers and researchers have been transformed into men of action dependent on political powers, participating in their decisions, working with enterprises, and accounting for the social aspirations of their fellow citizens. This socialization of scientific and technological activity was strongly accentuated during the twentieth century, but it was not new. The process of constructing purely scientific knowledge, however, was not impeded or detoured—quite the opposite.

The proximity of science, technology, and society underscores the importance of so-called hybrid institutions in constructing and applying knowledge. The French electrical laboratory created after the 1881 Paris exposition, the German Mechanisch-Physikalisches Institut founded in 1887, and the Institut de la Soudure begun in France in 1930 are outstanding examples. These institutions defined standards and organized markets to stimulate proper development of different disciplines, thereby reinforcing the proximity of technological and scientific learning. At national and global levels, they created methods to test quality. Many of them have functioned as centers of collective research.

In the end, though, it is the consumer who directly determines the orientation of technological research. The contention, popular among analysts of consumption, that consumers are completely subject to marketing strategies is no more credible than would be a notion of their absolute independence from marketing stimulus. The attention that promoters of technological change devote to consumers' needs and social demands is nothing new. It was evident in the early modern era, when technology and science were regarded as means of social reform and technological progress as an instrument of social betterment. Certainly, popular acceptance of this concept was soon offset by radical critiques concerning the nefarious effects of technological change on health, working conditions, and safety. Certain technologies like steam engines, chemicals, and metallurgy were particularly implicated by these criticisms, and justifiably so. But it was the utilization of technology, rather than technology itself, that came into question. It was widely held that new ways would open to correct these nuisances. Thus, electricity was regarded as a means to correct the faults of steam. These criticisms at least partially refocused research efforts in the 1960s without obviating majority approval for technological change. Environmental problems, ever more serious since the 1960s, and anxiety about the intrinsic dangers of certain technologies have finally led to initial efforts to reorient research, including basic research. But they have also provoked concern about global research's capacity to find appropriate responses to environmental problems.

As previously noted, the first industrial revolution was characterized by a wish to improve the daily life of consumers in all classes of society, that is, to create a consumer society. In the nineteenth century, the ideology of hygiene inspired the sanitation of cities and homes, as well as the improvement of many other aspects of social life such as public lighting and mobility. One of the essential factors in the conquest of a better future was the creation of new objects, to which inventors ascribed liberating virtues in the realms of mobility, family, media, tourism, or conviviality. These objects were viewed as an infinite expansion of the possibilities for self-realization. Their design evolved with engineers' informal perception of their imperfections and, quite early, through the use of polling and marketing that aimed to categorize consumers and thereby satisfy their tastes. Marketing managed to gain force without totally alienating individuals. The consumer could evaluate each product and thus influence its evolution toward often-unanticipated uses and forms. In this sense, technological change can be considered a social construction.

Large technological networks also emerged and developed in response to social expectations. Water networks were a reply to disastrous conditions in the lives of citizens who suffered from intolerable odors and repeated epidemics. Railways brought a solution to a demand for mobility that was not fully met by roads and canals. Telegraph and telephone networks met a desire to communicate and to gather information instantly. They were conceived to ensure the functioning of physical networks like electricity and the railway, and to assuage the glut of information that seemed to know no limit.

Once created, networks tended to expand and merge by passing from a regional to a national or international scale in an extraordinary variety of ways, although for some of them, such as railways, a trend toward dissolution has now appeared in Europe.

Five Answers

It remains, in view of this volume's prefatory remarks, to answer five questions in order to define the main components of technological change in historical perspective.

1. Seen over the long term, technological choices have an essentially cultural character. Engineers and scientists, whether they define the general orientation of technological change or make a particular choice during its development, do nothing more than transfer society's needs and aspirations into material terms. That was certainly the case for hygienics and the search for comfort and well-being in the nineteenth century, but also for mass consumption, transportation, and communications in the twentieth century. Once these choices became settled, they gave birth to often-unexpected cultural practices. As noted, the development of this civilization became increasingly difficult to master. Only intellectuals and researchers were able to correct the drift of culture and the environment by redefining long-term technological choices and relevant disciplines.
2. While the importance of individual inventors cannot be overlooked, the enterprise should be considered the primary site of innovation and technological change. The function of research went through a series of transformations in four stages: (1) in the artisanal workshop, innovation depended on the continual and spontaneous adaptation of tools and machines to their use; (2) in the business enterprise, the workshop was a meeting place for accumulated know-how and knowledge acquired elsewhere in schools, laboratories, or other workshops; (3) in the vertically integrated enterprise, research became one of the components of a multi-divisional and centralized organization; (4) in enterprises organized into a network, research activities took leave of the ivory tower in which centralized organizations had enclosed them and became variously integrated into a network of relationships and exchanges of knowledge among many entities and variables.
3. The interaction of trades within the artisanal world prefigures the vast network of exchanges that the enterprise creates with other enterprises, research laboratories, and actors in technological change. These networks have a multiform character: dialogue between customer and merchant or between user and manufacturer; integration of research programs and collective marketing; cooperation with nearby local or sectoral enterprises and with concerned social actors. They may create social groups uniting other enterprises and dependent or associated organizations in order to promote and develop an ensemble of new products and procedures. In such a context, competition among enterprises is no longer limited to a single confrontation between quality and price. It occurs in a much more complex social and political ambiance where the complementarity and solidarity established among various actors and members of the social group play a considerable role. This

is the result of anchoring an enterprise or large laboratory, as well as its teams and collaborators, in social and political life.
4. The construction of technological know-how is the result of an encounter among several types of knowledge. The know-how of artisans and workers is a product of a continual apprenticeship along the arc of experience within a given discipline. Through observation of artisanal practices, experts and engineers elaborate formalized knowledge. It may also result from the invention of new procedures or products through deepened knowledge acquired during the process of manufacturing and its dysfunctions. Little by little, specific new technological disciplines are created, such as chemical engineering. At the same time, engineers apply scientific principles learned in schools. Finally, the early creation of hybrid research centers and laboratories reinforced a proximity of technology and science that favors the formation of a corpus of knowledge within each discipline, combining various types of knowledge without suppressing their differences.
5. To conclude these analyses, it is evident that the concept of industrial revolution is valid, provided that it is integrated into a broad vision of successive episodes during two long periods: the twelfth century to the beginning of the nineteenth, and then the nineteenth century to the end of the twentieth. The medieval and early modern era from the twelfth century to the 1760s led thereafter to the first industrial revolution. On the basis of new technologies that appeared, a coherent technical system could be constructed between 1830 and 1880. Founded on the technologies of mechanics and steam power, it encountered obvious limits by 1870, particularly in the use of energy that experienced important dysfunctions. Responses to these problems were provided by the new technologies of the second industrial revolution from 1880 to 1900, which culminated in a complete recasting of the technological system, exemplified by electricity, the internal combustion engine, organic chemistry, and the initial progress of electronics. Meanwhile, the institutionalization of research made a leap forward.

Between 1900 and the 1960s the new technologies that appeared during the second industrial revolution extended to mass production and to technological networks on a grand scale. One of the major components of this evolution was the diffusion of electronics and computers. After 1970 and especially in the 1980s a radial transformation occurred, creating a society dominated by communications and attempts to resolve environmental problems. This development constitutes a third industrial revolution.

Notes

Part One: Introductory Text

1. Fernand Braudel, *Civilisation matérielle et capitalisme* (Paris, 1967), p. 251.
2. John R. Harris, *Essays in Industry and Technology in the Eighteenth Century: England and France* (Hampshire, 1992), p. 6.
3. Cited by Philippe Braunstein, "À l'origine des privilèges d'invention aux XIVe et XVe siècles," in François Caron (ed.), *Les Brevets. Leur utilisation en histoire des techniques et de l'économie* (Paris, 1985), p. 59.

Chapter 1: The Artisanal Mode of Knowledge

1. François Crouzet, *The First Industrialists: The Problem of Origins* (Cambridge, 1985), p. 25.
2. David S. Landes, *Revolution in Time: Clocks and the Making of the Modern World*, rev. ed. (Cambridge, MA, 2000), p. 223.
3. Chris Evans and Gören Rydén, "Kinship and the Transmission of Skills: The Bar Iron Production in Britain and Sweden, 1500–1860," in Maxine Berg and Kristine Bruland (eds.), *Technological Revolutions in Europe: Historical Perspectives* (Cheltenham, UK, and Northampton, MA, 1998), p. 188.
4. Ibid., p. 190.
5. Ibid.
6. Denis Woronoff, *L'industrie sidérurgique en France pendant la Révolution et l'Empire* (Paris, 1984), p. 59.
7. Landes, *Revolution in Time*, p. 220.
8. Bernard Carlson, "Building Edison's Laboratory at West Orange," *History of Technology* 13 (1991): 160.
9. Paul Feller and Fernand Tourret, *L'Outil*, 2nd ed. (Troyes, 2004), p. 39.
10. Harry M. Collins, "Expert Systems and the Science of Knowledge," in Wiebe E. Bijker, Thomas P. Hughes, and Trevor Pinch (eds.), *The Social Construction of Technological Systems: New Directions in the Sociology and History of Technology* (Cambridge MA, 1987), pp. 329–48.
11. See Karl Gunnar Persson, *Pre-Industrial Economic Growth: Social Organization and Technological Progress in Europe* (Oxford and New York, 1988).

12. Odette Chapelot and Jean Chapelot, "L'artisanat de la poterie et de la terre architecturale: un moyen de connaissance des sociétés rurales du Moyen Âge," in Mireille Mousnier (ed.), *L'Artisan au village dans l'Europe médiévale et moderne* (Toulouse, 2001), p. 119.
13. Feller and Tourret, *L'Outil*, p. 47.
14. Renate Eikelmann, "Les émaux sur ronde-bosse," in Elisabeth Taburet-Delahaye (ed.), *Paris 1400. Les arts sous Charles VI* (Paris, 2004), pp. 165–67.
15. Gérard Gayot, *Les Draps de Sedan, 1646–1870* (Paris, 1998), pp. 97–140.
16. Steven L. Kaplan, *Les Ventres de Paris. Pouvoir et approvisionnement dans la France d'Ancien Régime* (Paris, 1988), p. 189.
17. Yves Coutant, "Lexique et technique du moulin à vent destiné à la mouture du blé d'après les comptes flamands des XIVe et XVe siècles," thesis at the University of Lille III (1990), vol. 3, p. 516.
18. Cited by Stanley D. Chapman and Jonathan D. Chambers, *The Beginnings of Industrial Britain* (London, 1970), p. 16.
19. Ibid., p. 17.
20. Landes, *Revolution in Time*, p. 216.
21. Ibid., p. 296.
22. Liliane Hilaire-Pérez, *L'Invention technique au siècle des Lumières* (Paris, 2000), p. 137.
23. Jean Meuvret, *Les Ébénistes du XVIIIe siècle français* (Paris, 1965), pp. 15–16.
24. Harris, *Essays in Industry and Technology*, pp. 12, 31.
25. Braunstein, "À l'origine des privilèges d'invention aux XIVe et XVe siècles," pp. 53–60.
26. Corinne Maitte, "Corporation et politique au village: Altare entre migrations et différenciation sociale, XVIe–XIXe siècle," *Revue historique* 617 (2001): 53.

Chapter 2: From Artisan to Expert

1. Bert de Munk, "La reproduction d'une crise. Grands maîtres entrepreneurs chez les menuisiers d'Anvers," in Gérard Bodé and Philippe Marchand (eds.), "Formation professionnelle et apprentissage (XVIIIe–XXe siècles)," *Revue du Nord* 17 (2003): 43.
2. Jacqueline Senker and Wendy Faulkner, "Networks, Tacit Knowledge and Innovation," in Rod Coombs et al. (eds.), *Technological Collaboration: The Dynamics of Cooperation in Industrial Innovation* (Cheltenham, 1996), pp. 76–97.
3. Jean Le Rond d'Alembert, *Discours préliminaire de l'Encyclopédie* (Paris, 1965), p. 138.
4. Odette Chapelot (ed.), *Du projet au chantier. Maîtres d'ouvrage et maîtres d'oeuvre aux XIVe et XVIe siècles* (Paris, 2001), p. 115.
5. Philippe Contamine, *La guerre au Moyen Âge* (Paris, 1980), p. 333.

Chapter 3: Formalized Knowledge

1. Antoine Picon, *L'Invention de l'ingénieur moderne. L'École des ponts et chaussées, 1747–1851* (Paris, 1992), p. 221.
2. Ibid., p. 624.
3. André Chastel, *L'Art italien* (Paris, 1956), vol. 1, p. 194.

4. The full name was the Royal Society of London for the Improvement of Natural Knowledge.

5. Robert Halleux, "Chimistes provinciaux et révolution industrielle. Le cas de la Belgique," *Archives internationales d'histoire des sciences* 46 (1996): 12–22.

Chapter 4: Technological Adventures

1. See Akos Paulinyi, "Revolution and Technology," in Roy Porter and Mikulás Teich (eds.), *Revolution in History* (Cambridge, 1986), pp. 261–89.

2. See Richard L. Hills, *Power from Steam: A History of the Stationary Steam Engine* (Cambridge, 1989), p. 60.

3. See Trevor Griffiths, Philip Hunt, and Patrick O'Brien, "The Curious History and Imminent Demise of the Challenge and Response Model," in Berg and Bruland, *Technological Revolutions in Europe*, pp. 119–37.

4. Louis André, *Machines à papier. Innovation et transformations de l'industrie papetière en France, 1798–1860* (Paris, 1996), p. 79.

5. See Maxine Berg, *The Age of Manufactures: Industry, Innovations, and Work in Britain, 1700–1820* (Oxford and New York, 1985).

6. Cited by Berg, ibid., p. 272.

7. See Jean-François Belhoste, "Paris, carrefour de l'industrie et de la science mécaniciennes," *Cahiers d'histoire et de philosophie des sciences* 52 (Lyon, 2004), p. 221.

Chapter 5: Institutional Logic and the Dynamics of Knowledge

1. See Emmanuel Chadeau, *Le Rêve et la puissance. L'avion et son siècle* (Paris, 1996), and *De Blériot à Dassault. Histoire de l'industrie aéronautique en France (1900–1950)* (Paris, 1987).

2. Giovanni Dosi, *Technical Change and Industrial Transformation* (London, 1984), pp. 164–65.

3. Eugène-Oscar Lami (ed.), *Dictionnaire encyclopédique et biographique de l'industrie et des arts industriels*, cited by François Caron, *Les Deux révolutions industrielles du XXe siècle* (Paris, 1997), p. 55.

4. Nicole Chezeau, "Les débuts de la métallographie," *Sciences et techniques en perspective* 34 (1995): 79–102.

5. Ibid., p. 119.

6. Ibid.

7. Ibid., p. 67.

8. Ibid., p. 53.

9. Jürgen Kocka, "Entrepreneurs and Managers in the German Industrial Revolution," in Peter Mathias and M. M. Postan (eds.), *The Cambridge Economic History of Europe* (Cambridge, 1978), vol. 7, p. 571.

10. Cited in William Joseph Reader, *Imperial Chemical Industries: A History* (Oxford, 1925), vol. 2, p. 34.

11. See Gilbert Hottois, "Les techniciens dans la société," *Revue européenne des sciences sociales* 108 (1997): 47–59.

12. See James Albert Allen, *Studies in Innovation in the Steel and Chemical Industries* (Manchester, 1967).

13. Yves Lequin, "Le métier," in Pierre Nora (ed.), *Les Lieux de mémoire* (Paris, 1992), vol. 3, p. 388.

14. Jean-Pierre Daviet, "La Compagnie de Saint-Gobain de 1830 à 1939, une entreprise française à rayonnement international," thesis at the Sorbonne (Paris, 1983), vol. 5, p. 34.

15. Ibid., p. 335.

16. Ibid.

17. Karl Karmarsch, *Die Polytechnische Schule zu Hannover*, 2nd ed. (Hanover, 1856), p. 217, cited by Kees Gispen, *New Profession, Old Order: Engineers and German Society (1815–1914)* (Cambridge, 1989), p. 40.

18. Robert Angus Buchanan, *The Engineers: A History of the Engineering Profession in Britain, 1750–1914* (London, 1989), p. 162.

19. See Maurice Daumas, "Les mécaniciens autodidacts français et l'acquisition des techniques britanniques," in *L'Acquisition des techniques par les pays non initiateurs* (Pont-à-Mousson, 1973).

20. Buchanan, *The Engineers*, p. 162.

21. Peter Lundgreen, "Engineering Education in Europe and the U.S.A., 1750–1930: The Rise to Dominance of School Culture and the Engineering Professions," *Annals of Science* 47 (1990): p. 56.

22. Cited by Charles Rodney Ray, "Des ouvriers aux ingénieurs: le développement des écoles d'arts et métiers et le rôle des anciens élèves," *Les ingénieurs, culture technique* 12 (1984): 281.

23. Lundgreen, "Engineering Education in Europe and the U.S.A.," p. 54.

24. Ibid., p. 61.

25. See Gernot Böhme, "The 'Scientification' of Technology," in Wolfgang Krohn, Edwin T. Layton, Jr., and Peter Weingart (eds.), *The Dynamics of Science and Technology: Social Values, Technical Norms and Scientific Criteria in the Development of Knowledge* (Dordrecht and Boston, 1978), vol. 2, p. 237.

26. Jean-Pierre Williot, *Jules Petiet (1813–1871). Un grand ingénieur du XIXe siècle* (Paris, 2007), p. 56 (extract of a pamphlet distributed in 1829 to attract students).

27. Ibid., pp. 109–12.

28. Ibid., p. 109.

29. Ibid.

30. See Walter G. Vincenti, *What Engineers Know and How They Know It: Analytical Studies from Aeronautical History* (Baltimore, 1993).

31. Ibid., pp. 218–19.

32. Edward W. Stevens, Jr., *The Grammar of the Machine: Technical Literacy and Early Industrial Expansion in the United States* (New Haven, 1995), pp. 59–60.

33. Cited by Henry Marguenau, "Einstein's Conception of Reality," in Paul Arthur Schilpp (ed.), *Albert Einstein, Philosopher-Scientist* (Cambridge, 1970), p. 250.

34. See Bruno Belhoste and Konstantinos Chatzis, "L'enseignement de la mécanique appliquée en France au début du XIXe siècle," in Claudine Fontanon (ed.), *Histoire de la mécanique appliquée. Enseignement, recherche et pratiques mécaniciennes en France après 1880* (Paris, 1998), p. 29 (originally published in *Cahiers d'histoire et de philosophie des sciences* 46).

35. Olivier Darrigol, *Worlds of Flow: A History of Hydrodynamics from the Bernoullis to Prandtl* (Oxford, 2005), p. 138.

36. Bruno Belhoste and Louis Lemaître, "J.-V. Poncelet, les ingénieurs militaires et les roues et turbines hydrauliques," *Cahiers d'histoire des sciences et des techniques* 29 (1990): 14.
37. Ibid., p. 43.
38. Ibid., pp. 33–89.
39. Edwin T. Layton, Jr., "Millwrights and Engineers, Science, Social Roles, and the Evolution of the Turbine in America," in Krohn, Layton, and Weingart, *The Dynamics of Science and Technology*, pp. 67–69.
40. See Böhme, "The 'Scientification' of Technology," pp. 219–50.
41. Pap Ndiaye, *Du Nylon et des bombes. DuPont de Nemours, le marché et l'État américain, 1900–1970* (Paris, 2001), pp. 240–58.
42. Darrigol, *Worlds of Flow*, p. 324.
43. See Michel Letté, *Henri Le Chatelier (1850–1936), ou la science appliquée à l'industrie* (Rennes, 2004).
44. Chezeau, "Les débuts de la métallographie," p. 125.
45. Cited by Jeffrey Allan Johnson, *The Kaiser's Chemists: Science and Modernization in Imperial Germany* (Chapel Hill, 1990), pp. 14–15.
46. Dominique Pestre, "The Decision-Making Processes for the Main Particle Accelerators Built Throughout the World from the 1930s to the 1970s," *History and Technology* 9 (1992), cited by John Krige (ed.), *Choosing Big Technologies* (Philadelphia, 1993), pp. 164–65.

Chapter 6: Steam Engines

1. Jacques Payen, "Machines et turbines à vapeur," in Maurice Daumas (ed.), *Histoire générale des techniques* (Paris, 1978), vol. 4, p. 84.
2. Louis C. Hunter, *A History of Industrial Power in the United States, 1780–1930* (Charlottesville, VA, 1985), vol. 2, pp. 436–37.
3. Ibid., p. 250.
4. P. Debette, "Détente," in Charles Laboulaye (ed.), *Dictionnaire des arts et manufactures* (Paris, 1845), vol. 1.
5. William Ernest Dalby, *Valves and Valve Gear Mechanism* (1906), preface cited by Hunter, *A History of Industrial Power*, p. 145.
6. Cited by Hills, *Power from Steam*, p. 157.
7. Cited by Hunter, *A History of Industrial Power*, p. 39.
8. Ibid., p. 46.
9. Walter G. Vincenti, *What Engineers Know and How They Know It*, p. 219.
10. See Crosbie Smith, *The Science of Energy: Cultural History of Energy Physics in Victorian Britain* (Chicago and London, 1998).
11. Sadi Carnot, *Réflexions sur la puissance motrice du feu*, in a critical edition of Robert Fox (ed.) (Vrin, 1978), p. 43.
12. Muriel Guedj, "L'émergence du principe de conservation de l'énergie et la construction de la thermodynamique," thesis at the University of Paris-VII (2000), p. 25.
13. Cited by Crosbie Smith and M. Norton Wise, *Energy and Empire: A Biographical Study of Lord Kelvin* (Cambridge, 1989), p. 108.
14. Ibid., p. 243.
15. Ibid., p. 345.

16. Cited by Smith, *The Science of Energy*, p. 154.
17. Cited by Guedj, "L'émergence du principe de conservation," p. 145.
18. Cited by Smith and Wise, *Energy and Empire*, p. 88.
19. Cited by Smith, *The Science of Energy*, p. 151.
20. See Hunter, *A History of Industrial Power*, pp. 436–37.
21. Ibid., pp. 447–48.
22. Cited by Payen, "Machines et turbines à vapeur," p. 43.
23. See David F. Channell, "The Harmony of Theory and Practice: The Engineering Science of W. J. M. Rankine," *Technology and Culture* 23 (1982): 39–52.
24. Payen, "Machines et turbines à vapeur," p. 47.
25. Hunter, *A History of Industrial Power*, p. 448.
26. Ibid.
27. Ibid., p. 663.
28. Ibid., p. 666.
29. Ibid., p. 451.
30. Payan, "Machines et turbines à vapeur," p. 25.
31. Cited by Hunter, *A History of Industrial Power*, p. 252.
32. Ibid., p. 174.
33. Laboulaye (ed.), *Dictionnaire des arts et manufactures*, 5th ed. (Paris, 1881), vol. 2.

Chapter 7: The Chemical Industry

1. See Ndiaye, *Du Nylon et des bombes*.
2. Cited by John W. Servos, *Physical Chemistry from Ostwald to Pauling: The Making of a Science in America* (Princeton, 1996), p. 10.
3. Bernadette Bensaude-Vincent and Isabelle Stengers, *Histoire de la chimie* (Paris, 1993), p. 189.
4. Wilfred Vernon Farrar, "Science and the German University System, 1790–1850," in Maurice P. Crosland (ed.), *The Emergence of Science in Western Europe* (London, 1975), pp. 184–86.
5. Ibid., p. 157.
6. Bensaude-Vincent and Stengers, *Histoire de la chimie*, p. 193.
7. Robert Fox and Agusti Nieto-Galan (eds.), *Natural Dyestuffs and Industrial Culture in Europe, 1750–1880* (Canton, MA, 1999), p. xvi.
8. Anthony S. Travis, *The Rainbow Makers: The Origins of the Synthetic Dyestuffs Industry in Western Europe* (Bethlehem, PA, 1993), p. 233.
9. Ernst Homburg, "The Influence of Demand on the Emergence of the Dye Industry: The Roles of Chemists and Colourists," *Journal of the Society of Dyers and Colourists* 99 (1983): 329.
10. See Anne-Claire Déré, "Daniel August Rosenstiehl (1850–1916): An Alsatian Chemist in the Synthetic Dyestuffs Industry," in Ernst Homburg et al. (eds.), *The Chemical Industry in Europe, 1815–1914: Industrial Growth, Pollution, and Professionalization* (Dordrecht, 1998), p. 311.
11. Homburg, "The Influence of Demand on the Emergence of the Dye Industry," p. 329.
12. Cited by Travis, *The Rainbow Makers*, p. 35.
13. Ibid., p. 36.

14. Ibid., p. 68.
15. Ibid., p. 137.
16. Bensaude-Vincent and Stengers, *Histoire de la chimie*, p. 235.
17. Carsten Reinhardt, "An Instrument of Corporate Strategy: The Central Research Laboratory at BASF, 1868-1890," in Homburg et al., *The Chemical Industry in Europe*, p. 241.
18. Bensaude-Vincent and Stengers, *Histoire de la chimie*, p. 236.
19. Wolfgang Wimmer, "Innovation in the German Pharmaceutical Industry," in Homburg et al., *The Chemical Industry in Europe*, p. 285.
20. Reinhardt, "An Instrument of Corporate Strategy," p. 245.
21. Ibid., p. 258.
22. Ibid., p. 259.
23. Nicole Chezeau, *De la forge au laboratoire. Naissance de la métallurgie physique (1860-1914)* (Rennes, 2004), p. 158.
24. Servos, *Physical Chemistry from Ostwald to Pauling*, p. 40.
25. Werner Abelshauser et al. (eds.), *German Industry and Global Enterprise. BASF: The History of a Company* (Cambridge, 2004), p. 153.
26. Bensaude-Vincent and Stengers, *Histoire de la chimie*, p. 250.
27. Yasu Furukawa, *Inventing Polymer Science: Staudinger, Carothers, and the Emergence of Macromolecular Chemistry* (Philadelphia, 1998), p. 17.
28. Bensaude-Vincent and Stengers, *Histoire de la chimie*, p. 255.
29. Lionel Dumont, "L'industrie française du caoutchouc, 1828-1938: analyse d'un secteur de production," thesis at the University of Paris-VII (1997), p. 294.
30. Frank Greeway et al., "The Chemical Industry," in Trevor J. Williams (ed.), *A History of Technology*, vol. 6: *The Twentieth Century (1900-1950)* (Oxford and New York, 1978), p. 552.
31. Wiebe E. Bijker, "The Social Construction of Bakelite: Toward a Theory of Invention," in Bijker et al., *The Social Construction of Technological Systems*, p. 169.
32. Ibid., p. 176
33. Cited by Furukawa, *Inventing Polymer Science*, p. 118.
34. See Servos, *Physical Chemistry from Ostwald to Pauling*, p. 300.
35. Cited by Bijker, "The Social Construction of Bakelite," p. 167.
36. Cited by Furukawa, *Inventing Polymer Science*, p. 65.
37. Herman F. Mark, *From Small Organic Molecules to Large: A Century of Progress* (Washington, D.C., 1993), p. 25.
38. Ibid., p. 127.
39. Ibid., p. 78.
40. Responsible for this refusal was Fritz Haber, who was in charge of distributing funds for the acquisition of this type of expensive laboratory equipment. His relations with Staudinger had worsened during the war because of Staudinger's condemnation of chemical weapons.
41. Mark, *From Small Organic Molecules to Large*, p. 44.
42. Furukawa, *Inventing Polymer Science*, p. 78.
43. Ibid., p. 89.
44. Cited by David A. Hounshell and John Kenly Smith, *Science and Corporate Strategy: DuPont R&D, 1902-1980* (Cambridge, 1988), p. 223.
45. Ibid., p. 224.

46. Ibid., p. 225.
47. Furukawa, *Inventing Polymer Science*, p. 125.
48. Ibid., p. 130.
49. Ibid., p. 128.
50. Ibid., p. 138.
51. Ibid., p. 139.
52. Hounshell and Smith, *Science and Corporate Strategy*, pp. 239–40.
53. Furukawa, *Inventing Polymer Science*, p. 141.
54. Ibid., p. 185.
55. Cited by Sybil P. Parker (ed.), *McGraw-Hill Modern Scientists and Engineers* (New York, 1980), p. 18.

Chapter 8: Technological Interdependence and Consumer Needs

1. See Christopher Freeman, *The Economics of Industrial Innovation* (Harmondsworth, 1974), pp. 188–89.
2. See Wassily Leontief, *The Structure of the American Economy, 1919–1939: An Empirical Application of Equilibrium Analysis*, 3rd ed. (New York, 1960), p. 3. See also Ann P. Carter, *Structural Change in the American Economy* (Cambridge, MA, 1970); and Caron, "Histoire économique et dynamique des structures," *L'Année sociologique* 41 (1991): 107–28.
3. Nathan Rosenberg, "Technological Interdependence in the American Economy," *Technology and Culture* 20 (1979): 25–50.
4. Jean-François Belhoste et al. (eds.), *La Métallurgie comtoise (XVe–XIXe siècles). Études du Val de Saône* (Besançon, 1994), p. 44.
5. Jean-François Belhoste et al. (eds.), *La Métallurgie normande (XIIe–XVIIe siècles). La révolution du haut fourneau* (Caen, 1991), p. 57.
6. Report to stockholders by the executive council of the Compagnie du chemin de fer du Nord, June 1847.
7. Caron, *Histoire des chemins de fer en France*, vol. 1: *1740–1883*, 2nd ed. (Paris, 2005), p. 294.
8. Caron, *Histoire des chemins de fer en France*, vol. 2: *1883–1937* (Paris, 2005), p. 912.
9. See Eric von Hippel, *The Sources of Innovation* (New York and Oxford, 1987).
10. Janet T. Knoedeler, "Market Structure, Industrial Research, and Consumers of Innovation: Forging Backward Linkages to Research in the Turn-of-the-Century U.S. Steel Industry," *Business History Review* 67 (1993): 103.
11. Ibid., p. 105.
12. See Betty H. Pruitt and George D. Smith, "La Recherche chez Alcoa: les alliages aéronautiques par corrosion sous tension," *Cahiers d'histoire de l'aluminium* 6 (1990): 6–39.
13. Ibid., p. 14.
14. Ibid., p. 16.
15. Ibid., p. 34.

Chapter 9: Strategies and Social Networks

1. Philippe Mustar, "Les réseaux de distribution du gaz à la fin du XIXe siècle; construire la demande," in Fabienne Cardot (ed.), *L'Électricité et ses consommateurs* (Paris, 1987), pp. 203–17.

2. Christine Blondel, "Réponses d'une profession ancienne à de nouveaux besoins: les 'ingénieurs constructeurs' d'instruments électriques à la fin du XIXe siècle," *Bulletin d'histoire de l'électricité* 11 (1988): 113.

3. See Fabienne Cardot and François Caron (eds.), *Histoire générale de l'électricité en France*, vol. 1: *Espoirs et conquêtes, 1881–1918* (Paris, 1991), p. 35.

4. Christine Berthet and François Caron, "Réflexions à propos de l'exposition de Paris de 1881," *Bulletin d'histoire de l'électricité* 2 (1983): 9.

5. André Grelon, "La formation des ingénieurs électriciens," in Cardot and Caron, *Histoire générale de l'électricité en France*, p. 290. See also Éric Belloc, "La normalisation électrique en France," MA thesis at the University of Paris-IV (1985–86).

6. Bijker, "The Social Construction of Bakelite," p. 174.

7. Ibid., p. 176.

8. See Anne-Catherine Robert-Hauglustaine, "Le soudage des métaux en France. Un demi-siècle d'innovations techniques, 1892–1939," thesis at the École des Hautes Études en Sciences Sociales (1997), p. 143.

9. Tristan Gaston-Breton, "Georges Claude et L'Air Liquide," *Les Échos* (August 2003): 35.

10. Robert-Hauglustaine, "Le soudage des métaux en France," p. 94.

11. Ibid., p. 99.

12. Ibid., p. 294.

13. Ibid., p. 459.

14. The reflections of Alfred D. Chandler, Jr., have appeared in his numerous historical and theoretical works, notably *The Invisible Hand: The Managerial Revolution in American Business* (Cambridge, MA, 1977), and *Scale and Scope: The Dynamics of Industrial Capitalism* (Cambridge, MA, 1990).

15. Nathalie Greenan and Dominique Guellec, "Coordination within the Firm and Endogenous Growth," *Industrial and Corporate Change* 3 (1994): 194.

16. Ibid., p. 200.

17. Pierre Veltz, *Le Nouveau Monde industriel* (Paris, 2000), p. 178.

18. Ibid., p. 179.

19. Manuel Castells, *La Société en réseau*, vol. 1: *L'ère de l'information* (Paris, 1998), p. 202.

20. Ibid., p. 231.

21. See Hans Georg Gemünden, Peter Heydebreck, and Rainer Herden, "Technological Interweavement: A Means of Achieving Innovative Success," *R&D Management* 22 (1992): 359–76.

22. François Caron, *Les Deux Révolutions industrielles du XXe siècle* (Paris, 1997), p. 154.

23. Pierre Lanthier, "L'IGEC et l'organisation mondiale de l'industrie électronique dans l'entre-deux-guerres," in Dominique Barjot (ed.), *Vues nouvelles sur les cartels internationaux (1880–1980)* (Caen, 1994), p. 384.

24. See Marc C. Casson (ed.), *Global Research Strategy and International Competitiveness* (Oxford and New York, 1991).

25. Michel Callon, "Recherche et innovation en France: définition d'un cadre analytique," in Commissariat Générale du Plan, *Recherches et innovation: le temps des réseaux. Rapport du groupe "Recherche, technologie et compétivité"* (Paris, 1993), pp. 110–11.
26. Callon, "Sociologie des sciences et économie du changement technique: l'irrésistible montée des réseaux technico-économiques," in Centre de sociologie de l'innovation, *Ces réseaux que la raison ignore* (Paris, 1992), p. 54.
27. Ibid.
28. Ibid., p. 55.
29. Éric Robert, "Les relations université-industrie au sein des écoles d'ingénieurs grenobloises pendant la première moitié du XXe siècle," in Hervé Joly et al. (eds.), *Des barrages, des usines et des hommes. L'industrialisation des Alpes du Nord entre ressources locales et apports extérieurs* (Grenoble, 2002), p. 246.
30. Dominique Pestre, "Le CNRS, moyen d'une politique de la science," *Cahiers pour l'histoire du CNRS, 1939–1989* (Paris, 1989), p. 50.
31. Ibid.
32. Michel Bernardy de Sigoyer and Pierre Boisgontier, *Grains de technopole. Microentreprises grenobloises et nouveaux espaces productifs* (Grenoble, 1988), p. 110.
33. See Pierre-Éric Mounier-Kuhn, "L'informatique en France, de la Deuxième Guerre mondiale au Plan Calcul" (thesis at CNAM, 1999), vol. 1, p. 212.
34. Pestre, "Le CNRS, moyen d'une politique de la science," p. 156.
35. Ibid., p. 122.
36. Bernardy de Sigoyer and Boisgontier, *Grains de technopole*, p. 169.
37. Ibid., p. 32.
38. Giacomo Beccatini, "Les districts industriels," in Margaret Maruani et al. (eds.), *La Flexibilité en Italie*, cited by Florence Vidal, *Réseaux d'entreprises et territoire* (Paris, 2001), p. 83.
39. Alfred Marshall, *Principles of Economics* (1890), cited by Jean-Claude Daumas, "Le complexe technique de Roubaix-Tourcoing au XIXe siècle," in Michel Lescure (ed.), *La Mobilisation du territoire. Les districts industriels en Europe occidentale de XVIIe au XXe siècle* (Paris, 2006), p. 241.
40. Bruno Courault, "Les PME du 'monde choletais' (1945–2004)," in Lescure (ed.), *La Mobilisation du territoire*, p. 422.
41. Ibid., p. 417.
42. Ibid., p. 436.

Chapter 10: From Early Modern Times to the 1880s

1. See Gérard Jorland, *Une société à soigner. Hygiène et salubrité publiques en France au XIXe siècle* (Paris, 2010).
2. Pierre Arizzoli-Clémentel, "Les soieries de Lyon sous la Restauration et la monarchie de Juillet," in *Un âge d'or des arts décoratifs, 1814–1848* (Paris, 1991), pp. 22–23.
3. Henri Pansu, *Claude-Joseph Bonnet. Soierie et société à Lyon et en Bugey au XIXe siècle* (Lyon, 2003), p. 215.
4. Jules Michelet, *Le Peuple* (Brussels, 1846), p. 33.
5. Charles Laboulaye (ed.), *Dictionnaire des arts et manufactures* (5th ed., 1881), cited by Caron, *Le Résistible Déclin des sociétés industrielles* (Paris, 1985), p. 90.

6. Gérard Jacquemet, "Urbanisme parisien. La bataille du tout-à-l'égout à Paris à la fin du XIXe siècle," *Revue d'histoire moderne et contemporaine* 26 (1979): 517.
7. *Enquête sur l'exploitation et la construction des chemins de fer* (Paris, 1863).

Chapter 11: Technological Networks and Communications

1. "Rapport sur la Compagnie de l'Est," Archives of the Crédit Lyonnais, DEEF 34738.
2. Marcel Marion, *Histoire financière de la France depuis 1715*, vol. 6: *1876–1914, de la IIIe République jusqu'à la guerre* (Paris, 1931), p. 140.
3. Cited in Henri Bouchard, "Extraction et concassage mécanique du ballast," *Revue Générale des Chemins de Fer* (1911): 333.
4. Statement by a Belgian delegate at the International Congress of 1892.
5. Cited by Girolamo Ramunni, "La mise en place du système électrotechnique," in Cardot and Caron (eds.), *Histoire générale de l'électricité en France*, vol. 1, p. 323.
6. Henri Morsel, "La situation de la production de l'électricité en 1914," ibid., vol. 1, p. 738.
7. Alain Beltran, *La Ville-lumière et la fée électrique. L'énergie électricité dans la région parisienne: service publique et enterprises privées* (Paris, 2002), p. 308.
8. Christophe Bouneau, "Le transport d'énergie et l'interconnexion," in Maurice Lévy-Laboyer and Henri Morsel (eds.), *Histoire de l'électricité, 1919–1946*, vol. 2: *1919–1946* (Paris, 1994), p. 781.
9. Gaston Bonnefont, *Le Règne de l'électricité* (Tours, 1895), p. 210, cited by Alain Beltran and Patrice A. Carré, *La Fée et la Servante. La société française face à l'électricité, XIXe–XXe siècle* (Paris, 1991), p. 113.
10. Robert Mallet-Stevens, "L'échange moderne," *Design* 5 (1981): 12–29, cited by Beltran and Carré, *La Fée et la Servante*, p. 238.
11. Bernard Decaux, "Radiocommunications et électronique," in Maurice Daumas (ed.), *Histoire générale des techniques*, vol. 5, p. 276.

Chapter 12: From Microprocessors to the Internet

1. Francis Balle and Gérard Eymery, *Les Nouveaux Médias* (Paris, 1990), p. 57.
2. See Chai Yeh, *Handbook of Fiber Optics: Theory and Applications* (San Diego, 1990), p. 12.
3. Emmanuel Hoog, *Mémoire année zéro* (Paris, 2009), p. 146.
4. Ibid., p. 51.
5. Didier Lombard, *Le Village numérique mondial. La deuxième vie des réseaux* (Paris, 2007), p. 61.

Chapter 13: Information Technologies and Society

1. See Castells, *La Société en réseau*, vol. 1.
2. See Michel Volle, *e-économie* (Paris, 2000), p. 256.
3. Ibid., p. 257.

4. Castells, *La Société en réseau*, vol. 1, p. 199.

5. Jacques Perriault, "Le four à micro-onde ou la cuisine en parallèle," *Autrement* 3 (1992): 123.

6. See Mathieu Flonneau, *Les cultures du volant. Essai sur les mondes de l'automobilisme, XXe et XXIe siècles* (Paris, 2008).

7. Patrice Carré, "Radio, télévision, des richesses infinies 1899–1939," in Catherine Bertho (ed.), *Histoire des télécommunications en France* (Paris, 1984), p. 114.

8. See Nigel Calder (ed.), *L'Europe scientifique. Recherche et technologie dans vingt pays* (Maastricht, 1992), p. 177, cited by Caron, *Les Deux Révolutions industrielles du XXe siècle*, p. 425.

9. Hoog, *Mémoire année zéro*, p. 189.

10. Ibid., p. 157.

11. Ibid., p. 146.

12. Ibid.

13. Ibid., p. 119.

14. Ibid., p. 155.

15. Christian Huitema, *Et Dieu créa l'Internet* (Paris, 1996), p. 24.

16. Lombard, *Le Village numérique mondial*, p. 83.

17. Ibid., p. 60.

18. Hoog, *Mémoire année zéro*, p. 155.

Selected Bibliography

Abelshauser, Werner, Wolfgang von Hippel, Jeffrey Allan Johnson, and Raymond G. Stokes (eds.). *German Industry and Global Enterprise. BASF: The History of a Company*. Cambridge, 2004.
Allen, James Albert. *Studies in Innovation in the Steel and Chemical Industries*. Manchester, 1967.
André, Louis. *Machines à papier. Innovation et transformations de l'industrie papetière en France, 1798–1860*. Paris, 1996.
Balle, Francis, and Gérard Eymery. *Les Nouveaux Médias*. Paris, 1990.
Barjot, Dominique (ed.). *Vues nouvelles sur les cartels internationaux (1880–1980)*. Caen, 1994.
Belhoste, Jean-François, et al. *La Métallurgie normande (XIIe–XVIIe siècles). La révolution du haut fourneau*. Caen, 1991.
———. *La Métallurgie comtoise (XVe–XIXe siècles). Étude du Val de Saône*. Besançon, 1994.
Beltran, Alain. *La Ville-lumière et la fée électricité. L'énergie électrique dans la région parisienne: service public et enterprises privées*. Paris, 2002.
———, and Patrice A. Carré. *La Fée et la Servante. La société française face à l'électricité, XIXe–XXe siècle*. Paris, 1991.
Bensaude-Vincent, Bernadette, and Isabelle Stengers. *Histoire de la chimie*. Paris, 1993.
Berg, Maxine. *The Age of Manufactures: Industry, Innovations, and Work in Britain, 1700–1820*. Oxford and New York, 1985.
———, and Kristine Bruland (eds.). *Technological Revolutions in Europe: Historical Perspectives*. Cheltenham and Northampton, MA, 1998.
Bernardy de Sigoyer, Michel, and Pierre Boisgontier. *Grains de technopole. Micro-entreprises grenobloises et nouveaux espaces productifs*. Grenoble, 1988.
Bertho, Catherine (ed.). *Histoire des télécommunications en France*. Paris, 1984.
Bijker, Wiebe E., Thomas P. Hughes, and Trevor Pinch (eds.). *The Social Construction of Technological Systems: New Directions in the Sociology and History of Technology*. Cambridge, MA, 1987.

Bonnefont, Gaston. *Le Règne de l'électricité*. Tours, 1895.
Bouchard, Henri. *Extraction et concassage mécanique du ballast*. Paris, 1911.
Braudel, Fernand. *Civilisation matérielle et capitalisme*. Paris, 1967.
Buchanan, Robert Angus. *The Engineers: A History of the Engineering Profession in Britain, 1750–1914*. London, 1989.
Calder, Nigel (ed.). *L'Europe scientifique. Recherche et technologie dans vingt pays*. Maastricht, 1992.
Callon, Michel. "Recherche et innovation en France: définition d'un cadre analytique," in Commissariat Générale du Plan, *Recherches et innovation: le temps des réseaux. Rapport du groupe "Recherche, technologie et compétivité."* Paris, 1993.
———. "Sociologie des sciences et économie du changement technique: l'irrésistible montée des réseaux technico-économiques," in Centre de sociologie de l'innovation, *Ces réseaux que la raison ignore*. Paris, 1992.
Cardot, Fabienne (ed.). *L'Électricité et ses consommateurs*. Paris, 1987.
———, and François Caron (eds.). *Histoire générale de l'électricité en France*. Vol. 1: *Espoirs et conquêtes, 1881–1918*. Paris, 1991.
Caron, François. *Les Brevets. Leur utilisation en histoire des techniques et de l'économie*. Paris, 1985.
———. *Le Résistible Déclin des sociétés industrielles*. Paris, 1985.
———. *Les Deux Révolutions industrielles du XXe siècle*. Paris, 1997.
———. *Histoire des chemins de fer en France*. 2 vols. Paris 2005.
Carter, Ann P. *Structural Change in the American Economy*. Cambridge, MA, 1970.
Casson, Mark C. (ed.). *Global Research Strategy and International Competitiveness*. Oxford and New York, 1991.
Castells, Manuel. *La Société en réseau*. Vol. 1: *L'ère de l'information*. Paris, 1998.
Chadeau, Emmanuel. *De Blériot à Dassault. L'Industrie aéronautique en France, 1900–1950*. Paris, 1987.
Chandler, Alfred D., Jr. *The Invisible Hand: The Managerial Revolution in American Business*. Cambridge, MA, 1977.
———. *Scale and Scope: The Dynamics of Industrial Capitalism*. Cambridge, MA, 1990.
Chapelot, Odette (ed.). *Du projet au chantier. Maîtres d'ouvrages et maîtres d'oeuvres aux XIVe–XVIe siècles*. Paris, 2001.
Chapman, Stanley D., and Jonathan D. Chambers. *The Beginnings of Industrial Britain*. London, 1970.
Chastel, André. *L'Art italien*. Vol. 1. Paris, 1956.
Chezeau, Nicole. *De la forge au laboratoire. Naissance de la métallurgie physique (1860–1914)*. Rennes, 2004.
Contamine, Philippe. *La Guerre au Moyen Âge*. Paris, 1980.
Coombs, Rod, et al. *Technological Collaboration: The Dynamics of Cooperation in Industrial Innovation*. Cheltenham, 1996.
Coquery, Natacha, et al. (eds.). *Artisans, industrie. Nouvelles révolutions du Moyen Âge à nos jours*. Lyon, 2004.
Crosland, Maurice P. (ed.). *The Emergence of Science in Western Europe*. London, 1975.
Crouzet, François. *The First Industrialists: The Problem of Origins*. Cambridge, 1985.
Darrigol, Olivier. *Worlds of Flow: A History of Hydrodynamics from the Bernoullis to Prandtl*. Oxford, 2005.
Daumas, Jean-Claude (ed.), *Les Systèmes productifs dans l'Arc jurassien. Acteurs, pratiques et territories (XIXe-XXe siècles)*. Besançon, 2004.
Daumas, Maurice (ed.). *L'Acquisition des techniques par les pays non initiateurs*. Paris, 1973.

——— (ed.). *Histoire générale des techniques*. Vol. 4: *Les Techniques de la civilisation industrielle: énergie et matériaux* (Paris, 1978). Vol. 5: *Transformation, communication, facteur humain.* (Paris, 1979).
Dosi, Giovanni. *Technical Change and Industrial Transformation*. London, 1984.
Feller, Paul, and Fernand Tourret. *L'Outil*. Paris, 2004.
Ferguson, Eugene S. *Engineering and the Mind's Eye*. Cambridge, MA, 1992.
Flonneau, Mathieu. *Les Cultures du Volant. Essai sur les mondes de l'automobilisme, XXe et XXIe siècles*. Paris, 2008.
Foray, Dominique, and Jacques Mairesse (eds.). *Innovations et performances. Approches interdisciplinaires*. Paris, 1999.
Fox, Robert, and Agusti Nieto-Galan (eds.). *Natural Dyestuffs and Industrial Culture in Europe, 1750–1880*. Canton, MA, 1999.
Freeman, Christopher. *The Economics of Industrial Innovation*. Harmondsworth, 1974.
Furukawa, Yasu. *Inventing Polymer Science: Staudinger, Carothers, and the Emergence of Macromolecular Chemistry*. Philadelphia, 1998.
Gayot, Gérard. *Les Draps de Sedan, 1646–1870*. Paris, 1998.
Gispen, Kees. *New Profession, Old Order: Engineers and German Society (1815–1914)*. Cambridge, 1989.
Gras, Alain. *Grandeur et dépendance. Sociologie des macro-systèmes techniques*. Paris, 1993.
Harris, John R. *Essays in Industry and Technology in the Eighteenth Century: England and France*. Hampshire, 1992.
Hilaire-Pérez, Liliane. *L'Invention technique au siècle des Lumières*. Paris, 2000.
Hills, Richard L. *Power from Steam: A History of the Stationary Steam Engine*. Cambridge, 1989.
Hippel, Eric von. *The Sources of Innovation*. New York and Oxford, 1987.
Homburg, Ernst, et al. *The Chemical Industry in Europe, 1850–1914: Industrial Growth, Pollution, and Professionalization*. Dordrecht, 1998.
Hoog, Emmanuel. *Mémoire année zéro*. Paris, 2009.
Hounshell, David A. *From the American System of Manufacturing to Mass Production, 1800–1932*. Baltimore, 1984.
———, and John Kenly Smith. *Science and Corporate Strategy: DuPont R & D, 1902–1980*. Cambridge, 1988.
Huitema, Christian. *Et Dieu créa l'Internet*. Paris, 1996.
Hunter, Louis C. *A History of Industrial Power in the United States, 1780–1930*. Vol. 2: *Steam Power*. Charlottesville, VA, 1985.
Johnson, Jeffrey Allan. *The Kaiser's Chemists: Science and Modernization in Imperial Germany*. Chapel Hill, 1990.
Joly, Hervé, et al. (eds.). *Des barrages, des usines et des hommes. L'industrialisation des Alpes du Nord entre ressources locales et apports extérieurs*. Grenoble, 2002.
Jorland, Gérard. *Une société à soigner. Hygiène et salubrité publiques en France au XIXe siècle*. Paris, 2010.
Kaplan, Steven L. *Les Ventres de Paris. Pouvoir et approvisionnement dans la France d'Ancien Régime*. Paris, 1988.
Krige, John (ed.). *Choosing Big Technologies*. Philadelphia, 1993.
Krohn, Wolfgang, et al. (eds.). *The Dynamics of Sciences and Technology: Social Values, Technical Norms and Scientific Criteria in the Development of Knowledge*. Vol. 2. Dordrecht and Boston, 1978.
Landes, David S. *Clocks and the Making of the Modern World*. Rev. ed. Cambridge, MA, 2000.

Leontief, Wassily. *The Structure of the American Economy, 1919-1939: An Empirical Application of Equilibrium Analysis*. 3rd ed. New York, 1960.
Lescure, Michel (ed.). *La Mobilisation du territoire. Les districts industriels en Europe occidentale du XVIIe au XXe siècle*. Paris, 2006.
Letté, Michel. *Henri Le Chatelier (1850-1936), ou la Science appliquée à l'industrie*. Rennes, 2004.
Lévy-Leboyer, Maurice, and Henri Morsel (eds.). *Histoire de l'électricité en France*. Vol. 2: *1919-1946*. Paris, 1994.
Lombard, Didier. *Le Village numérique mondial. La deuxième vie des réseaux*. Paris, 2007.
Marion, Marcel. *Histoire financière de la France depuis 1715*. Vol. 6: *1876-1914, de la IIIe République jusqu'à la guerre*. Paris, 1931.
Mark, Herman F. *From Small Organic Molecules to Large: A Century of Progress*. Washington, D.C., 1993.
Mathias, Peter, and M. M. Postan (eds.). *The Cambridge Economic History of Europe*. Vol. 7: *The Industrial Economies: Capital, Labour, and Enterprise*. Cambridge, 1982.
Meuvret, Jean. *Les Ébénistes du XVIIIe siècle français*. Paris, 1965.
Mossman, Susan, and Peter J. T. Morris (eds.). *The Development of Plastics*. Cambridge, 1994.
Mousnier, Mireille (ed.). *L'Artisan au village dans l'Europe médiévale et moderne*. Toulouse, 2001.
Ndiaye, Pap. *Du Nylon et des bombes. DuPont de Nemours, le marché et l'État américain, 1900-1970*. Paris, 2001.
Negroponte, Nicholas. *L'Homme numérique*. Paris, 1995.
Nora, Pierre (ed.). *Les Lieux de mémoire*. Vols. 2 and 3. Paris, 1992.
Parker, Sybil P. (ed.). *McGraw-Hill Modern Scientists and Engineers*. New York, 1980.
Persson, Karl Gunnar. *Pre-Industrial Economic Growth: Social Organization and Technological Progress in Europe*. Oxford and New York, 1988.
Picon, Antoine. *L'Invention de l'ingénieur moderne. L'École des ponts et chaussées, 1747-1851*. Paris, 1992.
Porter, Roy, and Mikulás Teich (eds.). *Revolution in History*. Cambridge, 1986.
Reader, William Joseph. *Imperial Chemical Industries: A History*. Vol. 2: *The First Quarter-Century, 1926-52*. Oxford, 1975.
Reich, Leonard S. *The Making of American Industrial Research: Science and Business at GE and Bell (1876-1926)*. Cambridge, 1985.
Scardigli, Victor. *L'Europe des modes de vie*. Paris, 1987.
Schilpp, Paul Arthur (ed.). *Albert Einstein, Philosopher-Scientist*. Cambridge, 1970.
Schmidt, Susanne, and Raymond Werle. *Coordinating Technology: Studies in the International Standardization of Telecommunications*. Cambridge, MA, 1998.
Servos, John W. *Physical Chemistry from Ostwald to Pauling: The Making of a Science in America*. Princeton, 1996.
Siemens, Georg von. *L'Évolution de l'électrotechnique: histoire de la maison Siemens*. Vol. 1: *L'Époche de la libre entreprise, 1847-1910*. Colmar, 1965.
Smith, Crosbie. *The Science of Energy: Cultural History of Energy Physics in Victorian Britain*. Chicago and London, 1998.
———, and Matthew Norton Wise. *Energy and Empire: A Biographical Study of Lord Kelvin*. Cambridge, 1989.
Steffens, Henry John. *James Prescott Joule and the Concept of Energy*. Folkestone, 1979.
Stevens, Edward W. *The Grammar of the Machine: Technical Literacy and Early Industrial Expansion in the United States*. New Haven, 1995.
Taburet-Delahaye, Élisabeth (ed.). *Paris, 1400. Les arts sous Charles VI*. Paris, 2004.
Taccola, Mariano di Jacopo. *De Machinis: The Engineering Treatise*. 2 vols. Wiesbaden, 1971.

Travis, Anthony S. *The Rainbow Makers: The Origins of the Synthetic Dyestuffs Industry in Western Europe*. Bethlehem, PA, 1993.
Veltz, Pierre. *Le Nouveau Monde industriel*. Paris, 2000.
Vidal, Florence. *Réseaux d'entreprises et territoire*. Paris, 2001.
Vincenti, Walter G. *What Engineers Know and How They Know It: Analytical Studies from Aeronautical History*. Baltimore, 1993.
Virilio, Paul. *La Vitesse de circulation*. Paris, 1995.
Volle, Michel. *é-économie*. Paris, 2000.
Williams, Trevor I. (ed.). *A History of Technology*. Vol. 6: *The Twentieth Century (1900–1950)*. Oxford and New York, 1978.
Woronoff, Denis. *L'Industrie sidérurgique en France pendant la Révolution et l'Empire*. Paris, 1984.
Yeh, Chai. *Handbook of Fiber Optics: Theory and Applications*. San Diego, 1990.

Name Index

Agricola. *See* Bauer, Georg
Alembert, Jean Le Rond d' (1717–1783), 25–26, 64
Allen, John F. (1829–1900), 101, 103
Allen, Zachariah (1795–1882), 102–3
Alphand, Jean Charles (1817–1891), 188
Arkwright, Richard (1732–1792), 34, 44–45, 185
Armand, Louis (1905–1971), 194
Arrhenius, Svante (1859–1927), 119, 127
Arson, Alexandre, 187
Avogadro, Amedeo (1776–1856), 107

Bache, Alexander Dallas (1806–1867), 76
Bacon, Francis (1561–1626), 37
Baekeland, Leo Hendrik (1863–1944), 54, 124, 126, 128–29, 162–63
Baeyer, Adolf von (1835–1917), 110, 114, 116–17, 119, 122–23, 128, 130, 132
Bancroft, Wilder Dwight (1867–1953), 127, 133–34
Bauer, Georg (1494–1555), 26, 29, 36
Bayer, Friedrich (1825–1880), 114–15, 130
Becquerel, Alexandre Edmond (1820–1891), 161
Becquerel, Antoine César (1788–1878), 161
Bélidor, Bernard Forest de (1698–1761), 73
Bell, Alexander Graham (1847–1922), 54–55, 201
Belliss, George Edward (1838–1909), 101
Berlier, Jean-Baptiste (1841–1911), 190

Berliner, Emile (1851–1929), 200
Bernthsen, August (1855–1931), 117
Berthelot, Marcellin (1827–1907), 119, 127
Berthollet, Claude Louis (1748–1822), 40, 48, 119
Berry, Duke of (1340–1416), 22
Berzelius, Jöns Jacob (1779–1848), 106–8
Bessemer, Henry (1813–1898), 155–56
Bevan, Edward John (1836–1921), 125–26
Beylier, Charles, 173
Biringuccio, Vannoccio (1480–1540), 26
Birkeland, Kristian Olaf (1867–1917), 121
Biver, Ernest (1829–1889), 105
Biver, Hector (1824–1908), 64, 105
Black, Joseph (1729–1799), 39, 43
Boistel, Ernest, 196
Bolton, Elmer Keiser (1886–1968), 135–36
Bonaparte, Louis (1808–1873), 191
Bonaparte, Napoléon (1769–1821), 67, 188
Bonnet, Claude-Joseph (1786–1867), 184
Borda, Jean-Charles de (1733–1799), 39
Bosch, Carl (1874–1940), 121–22
Bouchard, Henri, 194
Bouchayer, Auguste (1867–1928), 173
Boulton, Matthew (1728–1809), 34, 43, 47, 85, 87–88
Bourdon, François (1797–1865), 153
Bousquet, Gaston du (1839–1910), 193
Bouton, Georges (1847–1938), 63
Boyden, Uriah Atherton (1804–1879), 75
Boyle, Robert (1627–1691), 36, 38

– 249 –

Breguet, Antoine (1851–1882), 161
Breguet, Louis Charles (1880–1955), 157
Breguet, Louis François Clément (1804–1883), 161
Brenier, Casimir (1832–1911), 173
Brunck, Heinrich von (1847–1911), 116
Brunelleschi, Filippo (1377–1446), 36
Bunsen, Robert Wilhelm (1811–1899), 120, 122
Burdin, Claude (1790–1873), 74

Cail, Jean-François (1804–1871), 65
Carcel, Bertrand Guillaume (1750–1812), 187
Carnot, Lazare (1753–1823), 39, 74
Carnot, Nicolas Léonard Sadi (1796–1832), 42, 91–95
Caro, Heinrich (1834–1910), 109, 112, 113–17
Caro, Nikodem (1871–1935), 121
Carothers, Wallace Hume (1896–1937), 60, 133–39
Cartwright, Edmund (1743–1823), 45
Cavé, François (1794–1875), 65, 85
Charpy, Georges (1865–1945), 57
Charrière, Eugène (1805–1885), 53
Cheape, William (1717–1806), 45
Chevreul, Michel Eugène (1786–1889), 109–10
Christofle, Charles (1805–1863), 53, 184
Clapeyron, Émile (1799–1864), 91–92
Claude, Georges (1870–1960), 54, 164
Clausius, Rudolf (1822–1888), 91–93, 95, 97, 120
Clément-Desormes, Nicolas (1778–1841), 56
Cochery, Adolphe Louis (1819–1900), 161
Coeur, Jacques (1400–1456), 21, 24, 28
Coffin, Donald, 136
Colbert, Jean-Baptiste (1619–1683), 7
Colladon, Jean-Daniel (1802–1893), 187
Collins, Arnold M., 135
Conant, James Bryant (1893–1978), 133
Condorcet, Antoine Caritat, Marquis de (1743–1794), 38–39
Cook, William, 186
Cordini, Antonio. See Sangallo, Antonio da
Corliss, George Henry (1817–1888), 101–3
Cort, Henry (1740–1800), 152–53
Couche, Charles-Henri-François (1815–1879), 154
Crompton, Samuel (1887–1954), 44

Cross, Charles Frederick (1855–1935), 125–26, 138
Crum, Walter (1796–1867), 111
Cusa, Nicholas of (1401–1464), 36

Dalby, William Ernest (1862–1936), 86
Dalton, John (1766–1844), 106–7
Darby, Abraham (1678–1717), 19–20, 152
Davis, George Edward (1850–1907), 77
De Forest, Lee (1873–1961), 201
Debye, Peter Joseph W. (1884–1966), 137
Delorme, Paul, 164
Deprez, Marcel (1843–1918), 196
Deutsch de la Meurthe, Henri (1846–1919), 162
Dickson, James Tennant, 61
Diderot, Denis (1713–1784), 26
Dreyfus, Camille (1878–1956), 126
Dreyfus, Henry (1882–1944), 126
Duhamel du Monceau, Henri Louis (1700–1782), 38, 48
Duhem, Pierre (1861–1916), 119–20
Duisberg, Friedrich Carl (1861–1935), 59, 115
Dumas, Jean-Baptiste (1800–1884), 107, 186–87
Durand, Louis (1837–1902), 112

Eastman, George (1854–1932), 125
Eaton, Amos (1776–1842), 68
Eddy, Henry Turner (1844–1921), 77
Edison, Thomas Alva (1847–1931), 52, 54–55, 69, 196, 200
Edwards, Humphrey (?–1829), 88
Einstein, Albert (1879–1955), 72, 137–38
Elder, John (1824–1869), 52, 90, 94–95, 98
Estienne, Charles (1504–1564), 36
Euler, Leonhard (1707–1783), 74,
Evans, Oliver (1755–1819), 42, 47, 88–89

Fairbairn, William (1789–1874), 47, 88
Farcot, Joseph (1823–1908), 85
Farcot, Marie-Joseph-Denis (1798–1875), 65, 101–2
Farcot, Marie-Joseph-Denis (1824–1908), 101–2
Fawcett, Eric William, 138
Fayol, Henri (1841–1925), 57
Ferranti, Sebastian Ziani de (1864–1930), 100
Ferraris, Galileo (1847–1897), 196

Fikentscher, Hans, 138
Fischer, Emil Hermann (1852–1919), 82, 128, 134
Flachat, Eugène (1802–1873), 70, 153
Fleming, John Ambrose (1849–1945), 201
Flint, Charles Ranlett (1850–1934), 202
Flory, Paul J. (1910–1985), 137
Fouché, Georges (1861–1931), 164
Fourier, Joseph (1768–1830), 91, 94–95
Fourneyron, Benoît (1802–1867), 42, 73–75
Francis, James Bicheno (1815–1892), 42, 73–76
Franck, Adolph (1834–1916), 121
Frary, Francis Cowles (1884–1970), 158
Freeth, Francis Arthur (1884–1970), 138
Freudenberg, Karl (1886–1983), 137
Fritschi, Jakob (1875–1940), 130
Froment, Gustave (1815–1865), 161

Galilei, Galileo (1564–1642), 18, 38
Gallier, Frédéric, 164
Gallois, Louis de (1775–1825), 53
Gates, Bill (1955–), 206
Gaulard, Lucien (1850–1888), 195–96
Gay-Lussac, Louis Joseph (1778–1850), 64, 81, 92–93, 107, 186
Gerber-Keller, Jean (1809–1884), 113
Gerhardt, Charles Frédéric (1816–1856), 107
Gerin, Gaston (1889–1943), 167
Gibbs, John Dixon, 196
Gibbs, Josiah Willard (1839–1903), 119–20
Gibson, Reginald Oswald (1902–1983), 138
Giffard, Pierre (1853–1922), 162
Gilchrist, Percy Carlyle (1851–1935), 57, 155
Girard, Charles, 112
Goodwin, Hannibal (1822–1900), 125
Goodyear, Charles (1800–1860), 123
Gordon, Lewis (1815–1876), 90, 94
Grace-Calvert, Frederick (1819–1873), 110–11
Graebe, Carl (1841–1927), 114
Graham, Thomas (1805–1869), 127
Gramme, Zénobe (1826–1901), 54, 161
Granjon, Raphaël (1878–1942), 165
Gray, Elisha (1835–1901), 201
Greene, Benjamin F. (1817–1895), 68
Griess, Johann Peter (1829–1888), 113
Grose, Samuel (1791–1866), 88
Guillaume I de Namur (1324–1391), 151
Gutenberg, Johann (1400–1468), 17

Guth, Eugen (1905–1990), 137, 142

Haber, Fritz (1868–1934), 59, 82, 121–22, 131
Hargreaves, James (1710–1778), 44–45
Harries, Carl Dietrich (1866–1923), 128, 130
Haussmann, Georges Eugène (1809–1891), 188
Henri IV, King (1553–1610), 32, 188
Héroult, Paul Louis-Toussaint (1863–1914), 54
Hertz, Heinrich (1857–1894), 201
Hesse, Albert, 164
Hilaire de Chardonnet, Louis-Marie (1839–1924), 125
Hill, Julian Werner (1904–1996), 135
Hindley, Henry (1701–1771), 33
Hirn, Gustave-Adolphe (1815–1890), 97
Hoff, Ted (1937–), 205
Hofmann, August Wilhelm von (1818–1892), 66, 109–15
Hofmann, Fritz (1866–1956), 130
Hollerith, Hermann (1860–1929), 54, 202
Hornblower, Johathan (1753–1815), 87
Horstmann, August Friedrich (1842–1929), 119
Houdry, Eugène (1892–1962), 61, 143, 224
Huygens, Christiaan (1629–1695), 37–38
Hyatt, John Wesley (1837–1920), 124–25

Isherwood, Benjamin Franklin (1822–1915), 97

Jacquard, Joseph Marie (1752–1834), 185
Janet, Paul (1863–1937), 173
Joule, James Prescott (1818–1889), 91, 93–95
Junkers, Hugo (1859–1935), 157

Kaplan, Viktor (1876–1934), 76
Katz, J. R., 132
Kay, John (1704–1780), 44
Kekulé von Stradonitz, August (1829–1896), 107, 112, 130
Kelvin, William Thomson (1824–1907), 81
Kettering, Charles Franklin (1876–1958), 202
Kilby, Jack St. Clair (1923–2005), 202
Kirk, Alexander Carnegie (1830–1892), 99
Klatte, Fritz (1880–1934), 138
Kleeberg, Werner, 128, 165
Koechlin, Camille (1811–1890), 109
Koechlin, Nicolas (1781–1852), 111

Kraemer, Elmer O. (1897–?), 134
Kuhn, Werner (1899–1963), 137, 142
Kunztmann, Jean (1912–1992), 175

Laboulaye, Charles (1813–1886), 86, 102
Lagrange, Joseph Louis de (1736–1813), 72
Laire, Georges de (1836–1908), 112
Langmuir, Irving (1881–1957), 58, 134
Latrobe, Benjamin Henri (1764–1820), 88
Laue, Max von (1879–1960), 131
Laurens, Camille, 97
Laurent, Auguste (1807–1853), 107, 110
Lauth, Charles (1836–1913), 111
Lavoisier, Antoine Laurent de (1743–1794), 39, 104, 106, 153, 186
Lawrence, Ernest Orlando (1901–1958), 83
Lean, Joel (1749–1812), 88
Leblanc, Nicolas (1742–1806), 47–48
Le Chatelier, André (1831–1939), 155, 164–65
Le Chatelier, Henry Louis (1850–1936), 81, 122, 164–65
Lee, William (?–1610), 44
Leffel, James, 75
Lenoir, Étienne (1822–1900), 33, 187
Lhomond, Nicholas, 186
Liebermann, Carl Theodor (1842–1914), 114
Liebig, Justus von (1803–1873), 107, 109, 111
Linde, Carl von (1842–1934), 120
Little, Arthur Dehon (1863–1935), 77
Livinstein, Ivan (1845–1916), 114
Lodge, Oliver Joseph (1851–1940), 201
Louis XIV, King (1638–1715), 7, 28
Lucius, Eugen (1834–1903), 115
Lumière, Auguste (1862–1954), 125
Lumière, Louis (1864–1948), 125

MacNeill, John Benjamin (1793–1880), 94
Mahan, Dennis Hart (1802–1871), 76
Mallet-Stevens, Robert (1886–1945), 198
Marconi, Guglielmo (1874–1937), 201
Mariotte, Edme (1620–1684), 42, 92
Mark, Herman Francis (1895–1992), 130–33, 136–38, 142
Marnas, Jean-Aimé (1828–1908), 110–11
Martin, Pierre (1824–1915), 155
Martius, Carl (1794–1868), 113
Marx, Karl (1818–1883), 63
Marshall, Alfred (1842–1924), 177
Maudslay, Henry (1771–1831), 46, 185
Maxwell, James Clerk (1831–1879), 81, 201

Médicis, Catherine de, Queen (1519–1589), 188
Merlin, Paul-Louis (1882–1973), 167, 175–76
Meyer, Kurt Heinrich (1883–1952), 132, 136, 142
Michael, Arthur (1853–1942), 128
Michelet, Jules (1798–1874), 184
Michelin, André (1853–1931), 54
Michelin, Édouard (1859–1940), 54, 162
Miles, George W., 126
Moissan, Henri (1852–1907), 164
Moncel, Théodore du (1821–1884), 161
Monge, Gaspard (1746–1818), 32, 39, 72
Morin, Arthur (1795–1880), 74, 81
Morris, Ellwood, 75
Murdock, William (1754–1839), 85–87
Murray, Matthew (1765–1826), 85–86

Natta, Giulio (1903–1979), 139
Navier, Henri (1785–1836), 73, 76, 92
Néel, Louis (1904–2000), 174–75
Neilson, James Beaumont (1792–1865), 153
Nernst, Walther Hermann (1864–1941), 82, 120, 122
Neumann, John von (1903–1957), 203
Newcomen, Thomas (1663–1729), 33, 42–43, 87, 102
Newton, Isaac (1642–1727), 38, 119
Neyret, André, 173
Nichol, John Pringle (1804–1859), 96
Nichols, James Burton (1902–1995), 134
Nicholson, Edward (1827–1890), 112
Nieuwland, Julius Arthur (1878–1936), 135
Normand, Charles, 99
Noyce, Robert (1927–1990), 202

Orléans, Duke of (1703–1752), 48
Osmond, Floris (1849–1912), 57, 155
Ostwald, Wilhelm (1853–1932), 82, 119–21, 127, 131
Ostwald, Wolfgang (1883–1943), 127
Otto, Nikolaus (1832–1891), 187

Palissy, Bernard (1510–1589), 36
Parker, Austin, 42, 74
Parker, Zebulon, 42, 74
Parkes, Alexander (1813–1890), 124
Parodi, Hippolyte (1874–1968), 197
Parsons, Charles (1854–1931), 100
Pasteur, Louis (1822–1895), 190

Paul, Jean-Raoul (1869–1960), 197
Pelouze, Théophile Jules (1807–1867), 64, 187
Pelton, Lester Allan (1829–1908), 74, 76
Périer, Auguste-Charles, 188
Périer, Jacques-Constantin (1742–1818), 188
Perkin, William Henry (1838–1907), 112–14, 140
Perrin, Jean (1870–1942), 131
Perrin, Michael Willcox, 138
Perrot, Bernard (?–1709), 22
Persoz, Jean-François (1805–1868), 109, 111
Petit, Alexis Thérèse (1791–1820), 92
Petty, William (1623–1687), 37
Peugeot, Armand (1849–1915), 54
Picard, Charles (1872–1957), 164
Pickard, James, 43
Pikles, Samuel (1878–1962), 130
Pionchon, Joseph, 173
Polanyi, Michael (1891–1976), 131
Poncelet, Jean Victor (1788–1867), 39, 42, 72–74
Porter, Charles Talbot (1824–1910), 100, 103
Portevin, Albert (1880–1962), 165
Pouzin, Louis (1931–), 210
Prandtl, Ludwig (1875–1953), 76
Priestley, Joseph (1733–1804), 39
Pupin, Michael Idvorsky (1858–1935), 201

Rankine, William John Macquorn (1820–1872), 52, 66, 90–91, 93–97
Raoult, François-Marie (1830–1901), 127
Redtenbacher, Ferdinand Jakob (1809–1863), 69
Regnault, Henri Victor (1810–1878), 56, 91, 93–95, 138, 187
Régnier, Edme (1731–1825), 33
Renault, Louis (1877–1944), 54
Rennie, John (1761–1821), 70
Riche de Prony, Gaspard François (1755–1839), 39
Riggs, Arthur, 99
Robert, Nicolas Louis (1761–1828), 45
Roberts, Richard (1789–1864), 44–45
Roebuck, John (1718–1794), 43, 47
Rogers, William Barton (1804–1882), 68, 76
Römer, Olaus (1644–1710), 37
Rosenberg, Pierre, (1869–1954), 165
Rothschild, Alphonse de (1827–1905), 161
Routin, Georges, 173

Sacc, Frédéric (1819–1890), 111
Sainte-Claire Deville, Henri (1818–1881), 56, 81
Sallman, James, 88
Sangallo, Antonio da (1484–1546), 29
Sanmicheli, Michele (1484–1559), 29
Savery, Thomas (1650–1715), 32–33
Say, Jean-Baptiste (1767–1832), 184
Scheele, Carl Wilhelm (1742–1786), 48
Schlumberger, Jules-Albert (1804–1892), 111
Schönbein, Christian Friedrich (1799–1868), 124
Schönherr, Otto (1861–1926), 122
Schutzenberger, Paul (1829–1897), 109, 111, 125
Seguin, Marc (1786–1875), 153
Serres, Olivier de (1539–1619), 36
Sickels, Frederick Ellsworth (1819–1895), 102
Siemens, Werner von (1816–1892), 82, 155
Sims, James, 88
Sloan, Alfred P. (1875–1966), 217
Smeaton, John (1724–1792), 33, 42, 70, 85
Smith, Adam (1723–1790), 13
Soufflot, Jacques (1713–1780), 152
Soutif, Michel (1921–), 174
Spill, Daniel (1832–1887), 124
Sprague, Frank J. (1857–1934), 196
Stanley, William, Jr. (1858–1916), 196
Staudinger, Hermann (1881–1965), 130–34, 136–39, 142
Stehelin, Hieronymus (1825–1913), 85
Stenhouse, John (1809–1880), 111
Stevens, Robert Livingston (1787–1856), 88
Stine, Charles (1882–1954), 13, 34
Stokes, George Gabriel (1819–1903), 73
Sualem, Rennequin (1645–1708), 28
Svedberg, Theodor (1884–1971), 130–32, 134
Swinburne, James (1858–1958), 129

Taylor, Frederick Winslow (1856–1915), 77, 200
Taylor, Geoffrey Ingram (1886–1975), 76
Taylor, John, 34
Telford, Thomas (1757–1834), 70
Tesla, Nikola (1856–1943), 196
Thiele, Friedrich Karl Johannes (1865–1919), 130
Thomas, Léonce (1812–1870), 97
Thomas, Sidney Gilchrist (1850–1885), 57, 155

Thompson, Benjamin (1753–1814), 186
Thomson, Elihu (1853–1937), 58–59, 195
Thomson, James (1822–1892), 92, 94
Thomson, William (1824–1907), 52, 90–95
Thurston, Robert Henry (1839–1903), 68, 76, 96, 98
Tirard, Pierre Emmanuel (1827–1893), 190
Tredgold, Thomas (1788–1829), 89, 186
Trevithick, Richard (1771–1833), 85–88, 92

Van Rensselaer, Stephen (1764–1839), 68
Van't Hoff, Jacobus Henricus (1852–1911), 119–20, 127
Vandermonde, Alexandre-Théophile (1735–1796), 33
Vauban, Sebastien de (1633–1707), 32
Velázquez, Diego (1599–1660), 44
Verguin, François-Emmanuel (1806–1865), 111–12
Victoria, Queen (1819–1901), 112
Vinci, Leonardo da (1452–1519), 26, 30

Wanklyn, James Alfred (1834–1906), 113
Washborough, Matthew, 43
Watt, James (1736–1819), 43, 47, 85–88, 92, 102
Werth, Jean (1855–1928), 57
Weskott, Johann Friedrich (1821–1876), 115
Westinghouse, George (1846–1914), 54–55, 100, 196
Whinfield, John Rex (1901–1966), 61, 138–39
Whipple, Squire (1804–1888), 76
Wilkinson, John (1728–1808), 46, 102, 152, 185
Wilm, Alfred (1869–1937), 158
Wöhler, Friedrich (1800–1882), 107, 111
Wood, Charles (1702–1774), 152
Wood, John, 152
Woolf, Arthur (1766–1837), 85–89, 92, 97–98, 101
Worthington, Henry Rossiter (1817–1880), 99

Ziegler, Karl Waldemar (1898–1973), 139
Zsigmondy, Richard Adolf (1865–1929), 120

www.ingramcontent.com/pod-product-compliance
Lightning Source LLC
Chambersburg PA
CBHW072148100526
44589CB00015B/2137